INTRODUCING

Mammals

AUTHOR: Kerry G. Everitt

ILLUSTRATOR: Ben Shannon

Pembroke Publishers Limited

©1999 The Federation of Ontario Naturalists

Pembroke Publishers
538 Hood Road
Markham, Ontario L3R 3K9
www.pembrokepublishers.com

Distributed in the U.S. by Stenhouse Publishers
P.O. Box 360
York, Maine 03909
www.stenhouse.com

Canadian Cataloguing in Publication Data

Everitt, Kerry G.
    Introducing mammals

(Hands on nature)
At head of title: federation of Ontario Naturalists.
Includes index.
ISBN 1-55138-111-7

1. Mammals – Study and teaching (Elementary). I. Shannon, Ben.
II. Federation of Ontario Naturalists. III. Title. IV. Series.

QL706.4.E93 1999   372.3'57   C99-931400-9

Advisors: Lynn Short, teacher, Board of Education for the City of Etobicoke
            Dan Stuckey, Kortright Centre for Conservation
            Kelly Horvath, teacher, Thames Valley District School Board

Pembroke Publishers gratefully acknowledges the support of the Department of
Canadian Heritage

Editors: Kat Mototsune, Deborah Sherman
Design: John Zehethofer
Cover Photography: Ajay Photographics
Typesetting: Jay Tee Graphics Ltd.

This book was produced with the generous assistance of the McLean Foundation.

For further information on the Federation of Ontario Naturalists and how
you can become a member contact:

FEDERATION OF
Ontario Naturalists
355 Lesmill Road Don Mills, Ontario M3B 2W8 (416) 444 8419
e-mail: info@ontarionature.org web site: www.ontarionature.org

Printed and bound in Canada
9 8 7 6 5 4 3 2 1

Contents

STUDENT ACTIVITY SHEETS

FACT SHEETS
- Common Shrew
- Little Brown Bat
- Snowshoe Hare
- Gray Squirrel
- Porcupine
- Beluga
- Black Bear
- Canada Lynx
- White-Tailed Deer
- Virginia Opossum
- Red Fox
- Harp Seal

Introduction

Ask your students what they have in common with a mouse, a horse, a bat, a whale and a polar bear. They may be surprised to learn that, although they all look very different, eat different foods, and live in different areas, they all share common characteristics. People, mice, horses, bats, whales and polar bears are all mammals.

Imagine running as fast as a car on the highway, or sleeping away the cold months of winter, or even holding your breath for 45 minutes. These are just some of the things that other mammals can do. Although humans are mammals, we seem to know very little about some of the other mammals with which we share this planet.

Introducing Mammals looks at

- various North American mammals in their habitats,
- mammals' place in our ecosystem,
- how mammals mate, nest and nurture,
- what we can do to better understand mammals and ensure their survival.

An introductory backgrounder gives you the information needed for the concept of each chapter. The bulk of the chapter consists of a set of lessons, each based on several activities. Particular background is given to you at the beginning of each lesson. You may choose to do all the activities or pick out those that you feel are most appropriate for your classroom. Whichever choice you make, you will find that the activities provide the focus for the learning experience. In some cases activities have an accompanying Student Activity Sheet that can be reproduced and given to the students. If the students keep their Activity Sheets together, they will have a comprehensive portfolio of facts about mammals.

You will find a series of fascinating *Did you know?* facts throughout the lessons. In the form of questions, they can be used as research assignments or as trivia questions. Have students use them as springboards to research other amazing mammal facts. In cases where data such as length, height or weight are given, have students relate these measurements to everyday items. For example, if the blue whale can reach a length of 33 m and one school bus is 12.5 m long, how many school buses parked end-to-end would this equal?

At the back of the book are illustrated mammal Fact Sheets designed for independent student use:

1. Encourage students to make use of the information in the Fact Sheets whenever they are assigned a research project dealing with mammals.

2. Have students look up and define the words in the "Words to Learn" section of the Fact Sheets. Further in-depth research on these topics or concepts may be of interest to some students.

3. Have students create their own mammal fact sheets, using the same headings and following the form of the Fact Sheets at the back of the book.

Introducing Mammals provides students with some of the basic building blocks of natural history knowledge. However, the activities use a broad, cross-curriculum approach so that you can teach about nature outside a designated science period. Students will be asked to observe, to communicate, to use mathematical skills, and to manipulate materials and equipment. In particular, students will be asked to study a wide variety of mammals, and their special characteristics; to prepare oral and written presentations; to roleplay; and to draw and construct.

After all, respect for living things and interest in and care for the environment are attitudes that are part of every curriculum. They should permeate everyone's life. Humans, as just one **example** of the thousands of mammal species **on Earth,** need to understand the role that **mammals** play in the ecosystem.

Enjoy learning with your students about the **common** — and not so common — mammals that inhabit the world, and about what **makes** them special.

A large colorful poster, designed for classroom display, is also available. It can be particularly useful as a focus in a learning centre.

The Introducing Mammals poster was illustrated by Judie Shore.

CHAPTER ONE

Incredible Mammals

Ask your students to name some mammals. How many mammals did they see on their way to school today? How many mammals can they see right now? Why are mammals an important part of the food chain, and what can we learn about ourselves by studying other mammals?

Introducing Mammals

It is believed that mammals first appeared on Earth about 200 million years ago, and their population increased substantially until after dinosaurs disappeared, about 65 million years ago. Today, it is estimated that there are more than 5,000 different species of mammals in the world, and humans are just one of them.

What's the difference between an animal and a mammal? The word animal is a broad term used to describe insects, birds, reptiles, amphibians, fish and other creatures that are not plants. Mammals are a type of animal with a backbone. Although mammals vary greatly in size — from bats smaller than bumblebees to the enormous blue whale, bigger than any dinosaur — they all have three characteristics making them different from other animals.

- All mammals have fur or hair at some stage in their lives.
- All mammals are warm-blooded.
- All mother mammals suckle their young on milk.

Mammals of all kinds have fur and/or hair on their bodies. Many mammals — such as wolves, chipmunks, bears and skunks — have a thick, recognizable fur coat. People don't have fur; however, we do have hair covering a large portion of our bodies. Whales have smooth skin, but do have small amounts of hair, mainly around their mouths.

All mammals are warm blooded (or homoiothermic, from *homo* = same, and *thermic* = heat) which means that they are able to maintain a constant internal body temperature. Mammals are able to keep their body warm through metabolism, which is the break down and digestion of food, and so they are called endotherms.

Mammals, as a rule, give birth to live young and feed their young on milk formed in the mammary glands of females. It is the characteristic of suckling young that gives mammals their biological name, Mammalia.

Did you know?

- Mammals are mostly water — between 50% and 90%, depending on the species.

Activities

1. Before beginning this unit, have students name as many characteristics of mammals as possible. Record their answers. Upon completing the unit, ask them once again to name as many characteristics of mammals as they can. Compare the two lists to show the students how much they have learned about mammals and their habitats.

2. Distribute Student Activity Sheet #1. Have students put a check mark beside the mammals and an X beside non-mammals. Have students circle one feature of the non-mammals that makes them not a mammal.

3. Distribute Student Activity Sheet #2. Some of the rhyming words may have to be a bit "stretched" — but keep it loose and fun. Have students think of other rhyming mammals.

4. Have students sit in a circle. The first student must come up with a mammal that begins with the letter A. After spelling the name of the mammal (with help if needed), the next person must name a mammal that begins with the last letter of the previous word (e.g., Antelope — Elephant — Tiger). Record the mammal names on the chalkboard. As a twist, have students name mammals in alphabetical order, rather than starting with the last letter of the previous mammal, to reinforce the alphabet and generate potential spelling words for future use.

5. Present the class with the following myths and ask students if they think the statements are true.

   Myth — A bat is a flying rodent.

   Fact — Although it is a mammal, like rodents, a bat is more closely related to you, according to taxonomic classification of mammals (see Lesson 2).

   Myth — Bears hibernate all winter.

   Fact — Some bears sleep more in the winter, but none are true hibernators.

   Myth — The opossum sleeps hanging upside down by its tail.

   Fact — The opossum sleeps lying down.

   Have students think of other myths about mammals and research the facts. Try playing a guessing game. Students can say either the fact or the myth and try to stump others in the class.

Species Classification

All living things have been given a scientific name, so each has the same name in every language. For example, the national mammal of Canada, the beaver, is known as *Castor canadensis* in the scientific community.

A classification system has been developed as a way to group living things according to characteristics, similarities and differences.

This method of grouping plants and animals is called taxonomy. The highest groups in the hierarchy of taxonomy contain many organisms that are only generally related. Moving to lower groupings leads to fewer members with more noticeably similar characteristics.

The classification of living things is divided into seven categories:

Kingdom, Phylum, Class, Order, Family, Genus, Species.

An example of the classification system used to divide living creatures can be adapted to objects that are more familiar to students. For example, Kingdom can be thought of as the province, Phylum is the city, Class is the neighbourhood, Order is the city block, Family is the building, Genus is the floor of the building, and Species can be the classroom you are in.

In order to further illustrate taxonomic classification, the following example is given for the classification of a domestic dog.

As you can see, each category gets more and more specific, with more common characteristics, until a species is made up of animals that most closely resemble one another. All mammals belong to the class Mammalia.

Did you know?

- Tooth and skull structure are used to classify mammals. They sometimes, but not always, determine what mammals eat, too. For example, a bear is omnivorous (eats plants and animal flesh) but is a member of the order Carnivora, primarily because of its tooth structure.

Primary Groupings	Taxonomic classification	Definition of taxonomic classification
Kingdom	Animalia	all animals
Phylum	Chordata	animals with spinal cords and hollow vertebrae (see Lesson 3)
Class	Mammalia	warm-blooded animals with hair, and feed milk to young
Order	Carnivora	mammals that primarily eat the flesh of other animals; with canine teeth specialized for tearing flesh; feet are plantigrade or digitigrade (see Lesson 5)
Family	Canidae	mammals with a varied diet; with incisors, canines and premolars specialized for eating flesh (see Lesson 10); feet are digitigrade (see Lesson 16) with nonretractable claws; limbs are designed for movement on the ground
Genus	*Canis*	dogs or dog-like mammals
Species	*familiaris*	members of the genus Canis domesticated by humans — the dog

Activities

1. Have students classify everyday things according to a certain set of criteria. For example, they can classify a miscellaneous assortment of things based on their color, shape, size, and what they are made of. Bring an assortment of plastic and metal hardware items into the class. Have students create a classification system to divide these items into groups (plastic vs. metal; hinged or not; with and without a screw head; straight and curved objects; etc.). What happens to the classification system if there is an object that is metal but is covered with plastic? What if an object is partially curved but also partially straight? How many sub-categories do you have to make in order to put all the pieces in relevant categories? Have students classify other things that they are familiar with; toys, CDs, music, etc.

2. Distribute Student Activity Sheet #3. Have students think of the characteristics of mammals that distinguish them from other classes of animals. Discuss what characteristics are unique to mammals by giving examples of animals that are mammals and those that are not (i.e., birds, reptiles, amphibians, insects, etc.).

Vertebrates

All animals on Earth can be divided into two groups, invertebrates and vertebrates. Animals without a backbone are invertebrates. Insects, spiders, worms and snails are just some of the millions of invertebrates found on Earth. Scientists estimate that 95% of all known animals are invertebrates.

Animals such as mammals, fish, birds, reptiles and amphibians have backbones running through their bodies. These animals are called vertebrates. Vertebrates first appeared on Earth over 400 million years ago.

All mammals have an internal skeleton, as compared to insects with exoskeletons (skeletons on the outside of the body). The skeleton is simply a frame on which the body is built. Before birth, the skeleton is not made of bone, but cartilage, the same flexible material that forms your nose and ears. Although cartilage is strong and flexible, it is not strong enough to support the weight of the mammal. As the mammal grows, the cartilage is reinforced by mineral salts. Over time, this deposition of minerals turns the cartilage into bone. In humans, this process takes about 20 years and, in the end, the average person's skeleton will weigh about 5 kilograms (11 pounds).

Under the skin, a tail is a continuation of the spine. All mammals have a tail, even humans. Although we cannot see our tails, they exist as four or five fused vertebrae at the base of our spine, called the coccyx (pronounced *cox-six*), also called the tail bone. There are many different types of tails and they all perform different functions. A beaver uses its long, flat tail as a rudder for steering in the water, or for

Submission   Imposing Attitude   Certain Threat   Self Confident

How a wolf communicates with its tail

slapping quickly against the surface of the water to warn of danger. A wolf's tail can tell you (and other wolves) how it is feeling. A whale uses its muscular tail to propel itself through the water. The flattened tail of the flying squirrel acts as a rudder in the air to help the animal glide — it also acts as a brake.

Activities

1. Examine bones from a chicken or turkey. Although these creatures are not mammals and have different bone structures, looking at the cartilage found between the breastbone and the bones themselves will help students understand the differences between these two substances. Students can also compare bird bones to mammal bones by looking at spareribs and steak bones. Ask students to think about how and why bones from mammals are different from bones found in birds.

2. Using Student Activity Sheet #4, examine the tails of different mammals. Try to identify the mammal that each tail belongs to and suggest reasons why it looks the way it does. What can the tail of a mammal tell you about how it is feeling? How can people let others know how they feel without talking?

3. Using Student Activity Sheet #5, cut out the bone shapes and build a mammal's skeleton by gluing the shapes onto bristol board. Be sure that students use the diagram of a real skeleton when putting together the "bones." Have students use Plasticine for tendons and cartilage to hold the "bones" together. Pieces of clear plastic wrap can be used to cover the skeleton and act as muscles. Tissue paper could then be placed on top to represent the skin of the mammal.

LESSON 4

Fur and Hair

Most mammals have two types of fur. The dense, soft layer closest to their bodies is known as underhair. This layer protects the animals from extremes of temperature by providing insulation against the heat and cold. Over the underhair is a layer of thicker, coarse hair known as the guard hairs. This outer coating protects the soft undercoat and gives the animal its color and patterning.

Hair grows out a hair follicle in the skin; each hair has a tiny muscle attached to it. These muscles allow the hair to raise and lower which adjusts airflow near the skin, and as a result, adjusts the animal's temperature. These

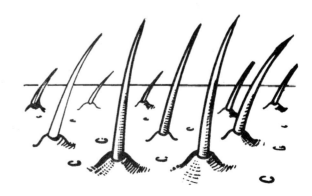

are the muscles that cause the hair on your arms to raise when you get cold, giving you goosebumps. Getting goosebumps is a way your body stays warm. Hair traps air close to the body to help regulate heating and cooling. A mammal's hair can also help repel wind and rain, insulating against the forces of weather.

Sebum is an oily substance that coats and waterproofs the hair of many mammals. This oil is especially important for mammals that spend a large portion of their lives in water. Beavers, muskrats and otters all have oiled fur to keep them warm and dry.

Most mammals undergo seasonal molts, when their fur is replaced very rapidly. In North America molting often occurs in the spring when the heavy winter undercoat is no longer needed. Humans do not molt; instead the hair on our bodies is continually replaced.

Some mammals use their hair as a form of defense. Raising their hair makes them look bigger and more aggressive, scaring predators. The quills of a porcupine and spines of a hedgehog are specialized hairs that help these creatures defend themselves against predators.

Did you know?

- Hair is made of keratin, a protein that also forms claws, bird feathers and horns. Hair that is growing out of the skin is dead. This is why it doesn't hurt to have yours cut.
- Some mammals, like cats, have special long, coarse hairs located around the nose and snout, called whiskers or vibrissae. They are used to increase the sense of touch.

Activities

1. Visit an outdoor education centre to speak with a trained naturalist about the importance of hair to mammals. Or try to obtain fur pelts from a local nature centre, museum, zoo or outdoor education facility. Second-hand stores may also have old fur pieces that students can examine. Have students examine the furs to learn about the different types of hairs that mammals have.

2. Collect hair samples from mammals for closer examination. Ask students to bring in a few hairs from their pets (dogs, cats, hamsters, gerbils, rabbits, etc.). If you go on a nature walk, see if you can find wild mammal hair. Pluck a few hairs from a person's head. Examine the hairs and try to discover similarities and differences. Look closer at the hair using a magnifying glass or a microscope.

CHAPTER TWO

Food for Thought

All animals have to eat to survive. Mammals, from the tiny meadow vole to the massive blue whale, and what they eat make up food chains and food webs. Some are predators, some are prey, and still others help get rid of dead plant and animal materials. Humans, along with all mammals, are an integral part of the circle of life.

LESSON 5

What's for Dinner

Plants are the first and lowest level of food chains. Plants are able to create their own food using the sun's energy. They are producers, because they produce their own food using non-living matter. Insects, reptiles, amphibians, birds and mammals are not able to produce their own food from the sun. These organisms are consumers, since they must eat other organisims in order to gain energy.

Organisms that eat only plant material such as grasses, shrubs and trees are called herbivores. Herbivores have special digestive systems that are able to break-down cellulose, the hard material found in plants. Herbivorous mammals must consume large quantities of plant material to obtain sufficient nutrients.

Grazing mammals such as cows and deer have multi-chambered stomachs to digest plant material. One of these stomachs is called the rumen, giving these animals the name ruminants. Microorganisms in their multiple stomachs assist in the break-down of cellulose and other plant tissues. The animal regurgitates the plant material, chews it again as cud, and swallows so that it is passed into another stomach chamber. Once passing through the reticulum, omasum and abomasum, the food is finally digested.

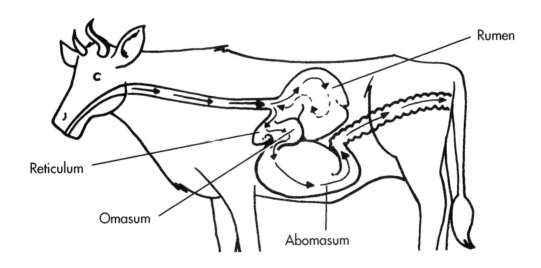

Rumen

Reticulum

Omasum

Abomasum

Species that eat other animals are carnivores. Insectivores such as shrews, moles and hedgehogs hunt insects and other small creatures. Although these mammals have poor eyesight, they use their keen sense of smell to locate their prey. Carnivores such as wolves and bobcats are members of the order Carnivora, with teeth well adapted for cutting and gripping flesh. These animals do not need to eat as much or as often as plant-eaters since meat contains a higher concentration of nutrients than plants.

Most carnivores hunt and kill other animals to eat. Animals that feed on dead animal material, or carrion, are called scavengers. The Virginia opossum is sometimes considered a scavenger. They can be seen on the side of the highway, eating animals that have been killed by vehicles. Unfortunately, opossums are also often killed by vehicles while scavenging for food along the road.

Some animals eat both plant material and other animals (insects, birds, mammals, etc.), and so are called omnivores (from *omni* = all). These animals have adapted to natural changes in their habitat and are able to eat whatever is available to them. Mammals such as bears, raccoons and skunks are omnivores.

Did you know?

- Humans are omnivores. What kinds of plant matter do you eat? What kinds of animal matter do you eat?

Food Chains

Insects eat leaves, birds eat insects, snakes eat birds, and foxes eat snakes. This is just one example of a food chain. A food chain is created by a number of organisms linked together in a specific order, determined by the order in which they feed on each other.

As you follow the food energy from plants up a food chain, energy is lost at each stage. That's why herbivores need to eat more than omnivores or carnivores. All energy comes from the sun, and plants use the sun's energy to make their own nutrients. From there, the energy flow in an ecosystem — through herbivores, carnivores, and decomposers — recycles the nutrients through the food chain.

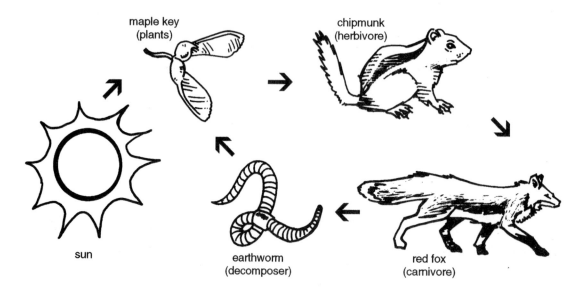

maple key (plants)

chipmunk (herbivore)

sun

earthworm (decomposer)

red fox (carnivore)

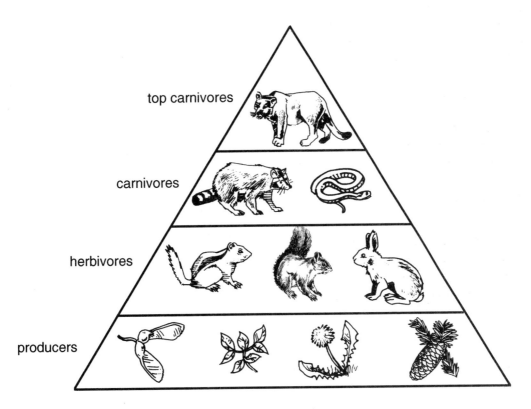

top carnivores

carnivores

herbivores

producers

Often neglected, but a very important part of all food chains, are the decomposers. Invertebrates such as earthworms, insect larva and carrion beetles feed on dead plant and animal material. Bacteria and fungi further breakdown dead plant and animal material and return these nutrients to the soil for the growth of plants; the plants, in turn, start the food chain cycle again. What would happen if there were no decomposers to return dead plant and animal matter to the soil?

In a balanced ecosystem, there must be more plants than herbivores, more herbivores than carnivores, and more lower carnivores than top carnivores. The above diagram of a food pyramid shows the relative number of producers, herbivores, carnivores and top carnivores in an ecosystem.

Did you know?

- Earthworms are important decomposers in the food chain, feeding on decaying plant and animal matter. They loosen and aerate the soil.
- Earthworms caused the burial of ancient ruins and can change human waste into fertile soil.
- More than one billion earthworms are collected each year for bait and research purposes.

Activities

1. Referring to the food pyramid diagram, ask students what they think would happen if there were more herbivores than plants, more carnivores than herbivores, and more top carnivores than carnivores. You may want to use a large triangle cut out of construction paper to visually illustrate the concept of a food pyramid. Turn the triangle upside down. What happens?

2. The song, "I Know an Old Lady Who Swallowed a Fly" is a good way to introduce the concept of a food chain. Find a copy on tape or CD, or distribute the lyrics and have the class sing. Challenge the students to come up with their own food chain songs, using this song as a model, or using another tune.

3. Have students make food chains.

   What you need:

   colored construction paper in a variety of colors
   scissors
   crayons, pencils, markers
   a piece of yellow bristol board
   tape or a stapler

   Cut colored construction paper into strips approximately 5 cm by 30 cm (2" x 12"). Use different colored paper to represent the different levels in the food chain. Print the names of the species that correspond to each color on the paper (younger children may want to draw the species) and link them together into a paper chain. Once the chains are finished, ask the children what is missing from the food chains; what do plants need to live? The sun should be added to each of the chains. Cut a circle out of the bristol board to represent the sun. Put on a bulletin board and extend the food chains outward like rays.

4. Have students create a food web.

   What you need:

   pictures of plants and wildlife — you can use Student Activity Sheet #6
   glue
   large pieces of construction paper or bristol board
   markers
   chalk
   tape/magnets

   With teacher assistance if necessary, have students use the pictures of plants and animals to create food chains of 4 or 5 organisms. See how many different chains you can make. Now glue the pictures onto the paper. Draw arrows between the organisms to show who eats who. Once you have many food chains completed, take a closer look. Can you see any relationships between the chains? Put the chains on the chalkboard. Draw lines between the species that are eaten or eat other organisms that are on different food chains, to make the food web. Ask students what would happen if one of the species were removed from the food web? Carefully erase lines that connect that species with others. Keep erasing lines between the food chains as the removal of one organism affects another.

5. Go outdoors and play the Food Chain Relay Game, to illustrate the loss of energy as you move up a food chain (the water represents the energy).

   What you need:

   2 pails or ice-cream buckets per group
   1 individual yogurt container per student
   water
   Student Activity Sheet #6
   — Punch small holes in the individual yogurt containers.
   — Distribute animal cards from Student Activity Sheet #6.
   — Have the students form groups to represent food chains, based on their animal cards. Note that each group must have the same number of students.
   — Have students line up in order: plant, herbivore, ominivore, carnivore (and decomposer if necessary). Put a pail full of water in front of the "plant" in each group, and an empty pail in front of the "carnivore" (or "decomposer").
   — The object of this relay is to transfer as much water from the plants to the pail at the end, passing the water from person to person using the yogurt containers. The group that ends up with the most water in the second pail (i.e. losing the least amount of energy), wins.

CHAPTER THREE

How Mammals Survive

Many mammals have special adaptations to allow them to survive and live in particular habitats: some have special hairs to keep them warm in cold weather; others have developed a "sixth sense" to cope with the darkness of night or the depth of water; many hibernate to survive the harsh winters of the north.

Keeping Warm and Keeping Cool

Unlike reptiles and amphibians, whose body temperature is influenced by the outside air, mammals control their internal body temperature from the food they eat. For this reason, mammals must eat more than most other animals. For example, a shrew must eat its body weight in food per day in order to maintain its high metabolic rate. That's like a student who weighs 45 kg (100 pounds) eating 437 small hamburgers or 2328 chicken nuggets every day. A pregnant female shrew will eat up to three times her weight every day!

Since mammals generate their own heat, they are able to remain active even in very cold conditions. To prevent heat loss, mammals have fur and hair, which act as insulators (see Lesson 4). Mammals such as beluga whales, without much fur or hair, have instead a thick layer of blubber or fat under their skin to keep them warm.

Mammals must also have a way to cool off in order to prevent damage to the body from overheating. Many mammals cool themselves by increasing the flow of blood near the surface of the skin. Heat in the blood is then able to more easily escape, cooling the animal.

Heat is also lost through evaporation from sweat glands. If the body becomes overheated, sweat glands secrete water onto the skin. The evaporation of this water helps cool the body. Some animals are not able to sweat and instead lick themselves so that their saliva evaporates. Covering the body with mud is another way mammals are able to cool off in the warm summer months. As the water in the mud evaporates, heat is drawn away from the animal. Mud also provides protection from the sun, a sort of natural sunblock. This cooling effect is the main reason why pigs wallow in the mud.

Members of the dog family (wolves, coyotes and foxes) cool themselves by panting. By sticking out their long, moist tongues and quickly drawing in air through their noses and mouths, these animals are able to cool themselves. This process causes a loss of moisture from the lungs, cooling the body from the inside out.

Polar bears become susceptible to overheating because of their thick fur and insulating fat. If a polar bear's body temperature begins to rise, it turns its face or rump into the wind. This is an example of conductive heat loss through those areas of the body where the skin is most exposed.

Did you know?

- It's important for all mammals to stay hydrated, especially in warm weather, to regulate temperature. How much water does a human need to consume each day to prevent dehydration?

- Polar bears and people have about the same normal body temperature, 37° Celsius (98.6 degrees Fahrenheit).

- The common shrew must eat continuously in order to maintain its body temperature. This tiny insect-eater can die if it goes more than four hours without eating!

Activities

1. Have students imagine they were shrews, and had to eat their weight in food every day. If a small McDonald's hamburger weighs 103 grams (4 ounces), how many hamburgers would each have to eat? One chicken McNugget weighs approximately 19 grams (.33 ounces). How many chicken nuggets would each student have to eat? Find the weights of other commonly eaten foods and see how many the students would have to consume.

2. To illustrate the concept of the cooling effect of evaporation, have the students try this simple experiment.

What you need:

2 thermometers
2 paper towels
fan

— Place the thermometers on a table. Record the temperature.
— Cover one with a regular paper towel and the other with a wet paper towel. Place a fan in front of the thermometers so it blows on both.
— Wait 5 minutes and record the temperature of each thermometer.
— After an additional 5 minutes, record again. Ask students what difference was observed, and why the wet thermometer had a lower temperature.

3. To further illustrate the concept of insulators, conduct this simple experiment.

What you need:

5 small glass jars with lids, e.g., large baby food jars
5 containers larger and taller than the jars, e.g., 750 mL (24 ounces) yogurt containers
measuring cup
water
5 thermometers
cotton balls or cotton batten
paper
wool (raw, or an old sweater or scarf)
shortening

— Cover the bottom of the large yogurt container with one of the insulators: wool, cotton, paper, shortening.
— Place a jar inside each yogurt container and lightly pack insulator around the sides and to the very top of the jar.
— Using the measuring cup, pour 250 mL (8 ounces) of hot tap water into each jar. Insert the thermometer and take the temperature reading after one minute. Record the temperature of the water, remove the thermometer and put the lid on the jar.
— Cover each jar with more insulator. Leave the jar for 5 minutes, then remove the lid and take the temperature once again. Do this again after 10 minutes, 15 minutes and 30 minutes. Record your findings for each of the jars.

Ask students the following questions: Which jar retained the most heat? Which jar lost the most heat the quickest? Which material was the best insulator? Why did we not put insulating material around one of the jars? (*Answer*: to show how quickly water looses heat under normal conditions.) Which types of mammals are represented by each of the insulators? (*Answer*: wool to represent sheep or other mammals with fluffy, thick fur; cotton to represent mammals with thinner coats, such as deer or opossum; paper to represent humans, who have skin but very little hair; and shortening to represent whales, seals, sea lions and polar bears who rely on blubber to keep warm.)

Hibernation

All mammals have different ways of dealing with the onset of cold winter weather. Some mammals grow a thick layer of fur and develop extra layers of fat to help keep themselves warm. Others protect themselves from very cold temperatures by sleeping through the winter, a process known as hibernation. During hibernation, which may last a few days or several weeks, the animal's heartbeat and breathing slows. By slowing down its body functions, the mammal conserves energy while it sleeps.

Think of how hungry you are when you wake up in the morning. A mammal that is hibernating may not eat for several weeks, but it doesn't get hungry as it sleeps. With slowed metabolism, the mammal can live off of its stored fat. Prior to entering hibernation, mammals must eat large quantities of food. The animal uses this food, stored as fat, for energy as it sleeps.

Eastern chipmunk   Black bear

Animals that hibernate have a substance called Hibernation Inducement Trigger (HIT) in their blood. This substance reacts to shorter days, change in temperature or a shortage of food. One study took a blood sample from a squirrel in the late fall, just prior to hibernation. HIT was detected in the blood sample. This sample was stored until the spring when it was injected into another squirrel. Although there was plenty of food, warm weather and longer days, the presence of HIT in the squirrel's blood caused it to go into hibernation.

Hibernating mammals have a special type of brown fat that is found across the back and shoulders, near the brain, heart and lungs. This fat sends a quick burst of energy to these vital organs when it is time for the animal to wake up in the spring.

Not all mammals who go to sleep in the winter are true hibernators. Some are light sleepers. True hibernators are deep-sleeping animals whose heartbeat, breathing and body temperature change drastically during the winter. Light sleepers do not undergo these drastic changes in body function.

True hibernators include: woodchucks, ground squirrels, and little brown bats.

Light sleepers include: eastern chipmunks, striped skunks, bears and raccoons

Did you know?

- A woodchuck, while hibernating, only breathes once every 5 minutes. During the summer they breathe 6-8 times a minute.
- When the little brown bat hibernates, its breathing reduces from 200 times per minute to as little as once every 5 minutes. It reduces its heart beat from 400-700 times per minute to only 7 times per minute.

Activity

1. As the days get shorter, there is less sunlight. Changes in day length trigger many mammals to begin hibernation. Ask students to research other changes in nature affected by the shortened hours of daylight. (Answer: bird and mammal migration, and leaves changing color in the fall).

Defense

Some animals will try to frighten away their enemies by using loud noises, showing and gnashing their teeth, hissing and barking or puffing up their fur to make themselves look larger and more ferocious.

Have you ever heard the expression "playing possum"? It comes from the Virginia opossum's unique way of warding off predators — it plays dead. Most predators are attracted to movement so, by lying still, the opossum does not draw attention to itself. Some scientists believe that opossums do not actually play dead but they become so frightened that they faint!

Some prey animals, such as muskoxen and bison or buffalo, show there is safety in numbers. When a predator is spotted, these animals often form a ring, with their heads facing outwards, in a formation to allow them to attack all at once. The young are often found in the centre of the ring where they are safest from predation. Animals such as rabbits, mice and deer will run when confronted, while others will attempt to fight back with horns, antlers, teeth or by kicking.

L E S S O N   9

Camouflage and Warning Colors

What would you wear if you wanted to hide in the ice and snow? In the grass? In the forest? You would want to wear colors similar to those found in nature, as do mammals that don't want to be seen. Camouflage is an important defense for many animals. By blending into their surroundings, mammals are able to hide from predators.

snowshoe hare   white-tailed fawn

Camouflage is especially important for young mammals unable to run away from predators. Deer are born with spotted fur unlike the solid brown coat of the adults. The spots help the fawn hide in the grass and low bushes on the forest floor. If the fawn had a solid-colored coat, it would stand out against the multi-colored background of the undergrowth. Spots are difficult for predators to see, and they may move on not even realizing that a meal is underfoot.

The arctic fox and snowshoe hare are well adapted to changes in their environment. These animals live in northern regions where there is a lot of ice and snow in the winter. Although both mammals are brown in the warm months, by the time winter arrives each has a snowy-white fur coat.

In the fall as the days get shorter, some mammals stop producing a pigment called melanin. Melanin is found in human skin, making some people's skin darker than others. This pigment also gives hair its color. In the winter, as the animal's hair is gradually replaced, the new hair lacks color and is white.

Thus, with only tiny bits of black showing, the arctic fox and snowshoe hare are well camouflaged in the snow. Camouflage makes it difficult for predators to find the hare, and makes it easier for the fox to sneak up on prey. In the spring, the lengthened days trigger the release of melanin and the fur once again turns color.

Some mammals have fur that camouflages them in their natural habitats. Other mammals, such as the skunk, have contrasting colors — not to hide but to stand out so that predators will recognize them and stay away. Certain colors and color-combinations tell other animals that these mammals are not to be bothered with.

Did you know?

- Decreasing daylight hours in the fall trigger the snowshoe hare's color change from brown to white.
- Some rabbits will freeze in place rather than run. By staying motionless, they may be invisible to the predator, who may lose interest.

Activities

1. A mammal's fur will often camouflage or hide it from view. Use Student Activity Sheet #7 to show students how camouflage is used by mammals to hide from predators. Circle the mammals.

2. To illustrate how certain colors are better for camouflage than others, go on a toothpick hunt.

   What you need:

   Colored toothpicks: 25 red, 25 blue, 25 green, 25 yellow, 25 orange and 25 plain toothpicks.

   — Hide colored and uncolored toothpicks in the schoolground before class starts.
   — Have students go outside and try to find the toothpicks. Do not tell them what colors they should be looking for or how many are hidden. Only allow them 5 to 10 minutes to find the toothpicks.
   — Have them meet back as a group and count the number of each color they have found. Most likely, they will have uncovered most of the brightly colored ones and few of the duller colors.
   — Instruct them to look for more and have them go outside once again. Once they have found as many as possible, have students explain why some were easier than others to find. How does this relate to camouflage in the wild?

3. The importance of camouflage to mammals can also be demonstrated in the classroom. Have students cut out two identical silhouettes of a mammal from a piece of newspaper (use a section with a lot of

printing on it). Glue one silhouette to a full piece of newspaper and the other to the colored comics or a store flyer. Students should be able to see the mammal shape hiding in the colored comics much easier than the one that is camouflaged in the black newspaper.

4. Think of some animals (mammals and non-mammals) that use color to warn potential enemies that they should stay away. (Possible answers: monarch butterflies, coral snakes, poison arrow frogs). Do we have warning colors in our every day lives? Think about school busses being bright yellow, shiny red fire trucks and red stop signs. How many others can students think of?

LESSON 10

Claws and Teeth

Claws and fingernails are made out of keratin — the same protein in hair and feathers. Most land mammals have claws that are used for defense, digging, hunting for food, climbing and/or holding things. Members of the cat family are able to retract their claws; however, most other mammals are unable to do this. Although people and apes do not have claws like other mammals, we do have nails that protect the sensitive tips of our fingers and toes.

For many mammals, it is their sharp teeth that allow them feed. Carnivores use their long, sharp canine teeth to catch and kill their prey. Teeth can also be used in defence, especially when an animal or its young is under attack. Mammals, such as deer and rabbits, that lack sharp teeth flee when danger approaches. Other mammals will bare their teeth, hoping the attacker will be frightened away. When necessary, most mammals will attack furiously with their teeth and claws.

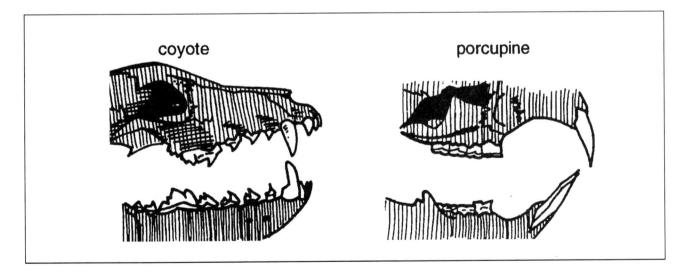

Comparing animal teeth

Mammals are the only animals that have two sets of teeth. The immature teeth fall out as the animal matures, to be later replaced by permanent teeth. Mammals have four groups of teeth in their jaws: incisors, canines, premolars and molars. Each of these types of teeth has different and important functions.

The carnivore teeth/jaw

Carnivores have sharp, pointed teeth in the front of their mouth. These incisors are used to scrape meat from the bones of their prey. Next to the incisors are the pointed canine teeth, used to hold and kill prey. If you have ever watched a vampire movie, the canine teeth are used to bite the necks of victims. At the back of a carnivore's mouth are the carnassial teeth, or molars. The top and bottom rows fit snugly together, to crunch bones and break down food into smaller pieces before swallowing. The jaws of meat-eating mammals move up and down, with limited side-to-side mobility.

Herbivore teeth/jaw

Plant-eating animals have incisors only in the front of their lower jaw. These teeth bite against a strong pad in the upper jaw, to cut through grasses and plant material. The back teeth are large and have flat tops, for grinding plant material before it is swallowed. Unlike carnivores, the jaws of herbivores move back and forth (side-to-side) to grind the plants. This continual grinding wears away the teeth and, in older individuals, the teeth may be quite smooth and ineffective.

Omnivore teeth/jaw

Mammals with a varied diet have a tooth and jaw structure that is between that of herbivores and carnivores. These mammals (such as humans and chimpanzees) can move their jaws both up and down and side-to-side. What benefits do you think there are to having the ability to vary your diet?

Rodents such as mice, rats and beavers have teeth that are different from herbivores and carnivores. The incisors of rodents grow throughout their lives. These creatures must

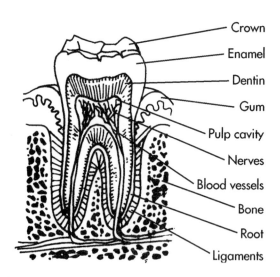

Crown
Enamel
Dentin
Gum
Pulp cavity
Nerves
Blood vessels
Bone
Root
Ligaments

continually gnaw on hard surfaces to prevent their teeth from overgrowing. If the teeth get too long, the rodent is unable to properly open its mouth and it will eventually die from starvation. Many rodents have a large space, called the diastema, between their incisors and molars. This gap helps a mammal such as the beaver hold onto twigs and branches. To prevent swallowing woodchips as they gnaw on trees and branches, beavers curl their lips into this space between their teeth.

Although different mammals have varying jaw and teeth arrangement, the structure of the teeth is very similar. The soft, inner portion of the tooth, containing nerves and blood, is the pulp. It is covered with dentine and then enamel.

Did you know?

- The mammal with the longest tooth is not a bear or a wild cat, it is a whale. The narwhal, also called the unicorn whale has one long tooth growing out of its mouth. This spiraling tooth can be up to 3 m (3 yards) in length!
- Enamel is the hardest substance found in a mammal's body!

Activities

1. Using Student Activity Sheet #8, have students try to match the tooth and jaw structures of mammals with the foods that they eat.

2. Bring mammal jaws and teeth into the classroom for students to examine. You may obtain specimens from a museum, outdoor education facility or nature centre, or use pictures of teeth and jaws from different mammals. Have students determine if the teeth or jaws are from a carnivore, herbivore or omnivore. How are the teeth different in size, structure, number? How does the structure of the jaw change depending on the animal's diet?

LESSON 11

Quills and Horns

Porcupines have quills, which are actually specialized hairs. Quills are usually 2 or 3 inches long and are very sharp. Despite what many people think, porcupines cannot throw or shoot their quills at enemies. When confronted by an enemy, porcupines will stamp the ground and swing their tails from side to side. This movement sometimes causes loose quills to fall to the ground. This loss of quills may be where the myth of porcupines throwing their quills originated. If attacked, however, the porcupine's quills may be released through muscle contraction. The quills must press against the attacker to be dislodged from the porcupine's skin. At the tip of each quill is a series of barbs, allowing the quill to easily puncture the skin of the attacker but not allowing easy removal. If not removed, the quills may go deeper and deeper into the flesh and, depending on their location, may eventually lead to death of the porcupine's attacker.

Horns and antlers are found only on herbivores, and are not needed for hunting or finding food. Instead, animals with horns and antlers use them to win contests — for the right to mate with a particular female, or the right to claim a good feeding ground. Some horned mammals will use their horns to protect their young, lowering their heads and attacking a predator with the sharp points.

Antlers are grown and shed each year. They are found on members of the deer family, such as deer, moose, elk, etc. These protrusions are made entirely of bone and do not have a protective covering like horns.

True horns are found on ruminants such as sheep and cows. They are not shed annually. These horns are made of specialized keratin (also found in claws, hair and feathers) which covers bone growing out of the animal's head. True horns are not branched and are found in both sexes. A rhinoceros horn is not a true horn but rather it is made out of hair that is cemented together into a horn-like structure.

Did you know?

- A porcupine's quills are actually a type of hair.
- Porcupines have an estimated 30,000 quills.
- You do not often find antlers in the forest because small rodents such as mice and chipmunks eat them. Antlers are rich in nutrients used by mammals — the rodents act as natural recyclers.
- The huge racks of antlers found on bull moose and deer are grown each year. As the animal gets older the antlers get larger.
- The more branched and pointed a deer's antlers are, the better and more nutritious its diet has been over the past year.

Activities

1. Distribute Student Activity Sheet #9. Have the students count the number of quills on the porcupine.
2. Use counting sticks and pretend they are quills. Make up math problems where students can use these quills for addition, subtraction, multiplication or division.

Example: A porcupine has 90 quills in its tail. A fox gets a little too close and the porcupine looses one third of its quills. How many quills does the porcupine still have? If seven quills fall to the ground but the rest lodge in the fox, how many quills does the poor fox have?

Chemical Defense

Skunks have a unique way of defending themselves. When confronted by a predator or other threat, a skunk will gnash its teeth, hiss and stamp the ground. If the perpetrator does not back off, the skunk will squirt an oily liquid at the attacker from a gland near its tail. This foul-smelling oil will discourage most enemies, as it also burns the eyes and nose of the creature that was sprayed.

Mammals are not usually poisonous, the way several types of snakes and insects are. One exception is the male duck-billed platypus, found in Australia. This mammal has spurs on its back ankles from which it can inject poison into an enemy. One Canadian mammal is venomous — the short-tailed shrew. This tiny rodent has a modified salivary gland that releases poison to flow though a hollow in the mammal's lower incisor. If bitten, a small animal may experience paralysis, respiratory failure and circulatory problems, all of which may eventually lead to death. The bite of a short-tailed shrew is not harmful to people but will cause pain and swelling near the bite.

Did you know?

- The Virginia opossum is one of the only animals that is immune to the venom of a rattlesnake.
- Skunks can spray their odorous oil up to 4 m (13 feet).
- The scent gland of a skunk holds 15 mL (0.5 ounces) of spray.
- Skunk oil can be smelled more than a kilometer (.6 miles) away.
- Great horned owls are one of the only creatures to be unaffected by a skunk's oily spray — owls do not have a sense of smell!

Activity

1. Divide the class in half. Have one group brainstorm "good" smells and the other "bad" smells. On a scale of 1 to 10, have students rate each smell by strength. Discuss whether good smells become bad if too strong.

CHAPTER FIVE

Getting Around

Ask students to think of different ways that mammals move through their environments. Unlike birds that fly and hop, and fish that swim, mammals have many different ways of moving. They can walk, run, pounce, leap, creep, dig, swim, hope, jump, and even fly! Once students have compiled a list, have them act out the movements — except for the flying, of course.

In the Air

Mammals that fly? Flying squirrels, despite their name, do not really fly — they glide. These mammals have a flap of skin between their front and back legs which, when held open, allows the squirrel to glide between branches of trees. The squirrel is able to steer itself by using its tail as a type of rudder.

The only mammal that truly flies is the bat. Bats have very long, thin fingers with stretched skin between them. This thin skin is also attached to the bat's ankles, making the wing. Despite the difference in sizes — the smallest bat is the size of a bumblebee and the largest has a wing span of 1.5 m (5 feet) — all bats are graceful and acrobatic fliers.

Did you know?

• The smallest mammal is the bumblebee or Kitti's hog-nosed bat, which lives in Thailand. This bat weighs no more than 2 grams (0.7 ounces) and is only 33 mm (1.3 inches) long.

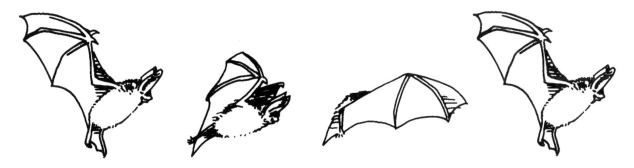

Bat in flight

In the Water

Ask your students to name as many creatures that live in the water as possible. No doubt they will name many fish, amphibians, reptiles, shelled creatures — and maybe a few mammals. Many students don't realize that some sea-creatures are mammals that breathe air, give birth to live young that suckle, and even have hair. Whales, seals and sea lions are all mammals living in or near the water.

Mammals that spend a lot of time in the water — such as beaver and muskrats — have very thick coats of fur to help stay warm. These mammals produce an oily substance which makes the mammal's fur waterproof, and repels water away from the body.

Did you know?

- The sperm whale can hold its breath for about 2 hours during deep dives.
- The largest creature to ever live on Earth is not a dinosaur — it's the blue whale! The largest specimen that has been found was a female weighing almost 200,000 kg (190 tons) with a length of more than 33 m (110 feet).
- The fastest marine mammal is the bull orca, which can reach speed of 55 km/h (34 mph).
- The most fatty diet is consumed by the polar bear during the spring and early summer, when it mainly feeds on recently-weaned baby seals which may be up to 50% fat.

Activities

1. Have students look at the bone structure of a whale on Student Activity Sheet #10. What similarities and differences can they see between this skeleton and the skeleton of other mammals? (Note the number of digits, backbone, skull, tail, etc.). What do these similarities suggest about the origins of whales? In what ways are whales different from fish?

2. You can illustrate how waterproofing works by explaining that water and oil do not mix. Fill a clear glass with water and add vegetable oil to show students this concept. You can replicate the oil on a mammal's fur by taking two pieces of felt and applying oil to one of them. With an eyedropper, place a single drop of water in the centre of the non-oiled felt. Now place the same amount of water on the oiled felt. Have the students observe the different results.

Underground

Many mammals live underground in burrows or dens. You'll find everything from the smallest mole to bears and wolves living underground. Although some mammals such as moles, groundhogs and badgers are proficient diggers and excavate burrows or dens in which to live and rest, others simply take over abandoned quarters.

Animals such as foxes will often only inhabit a den when it is time to have young. Babies are born underground where they are safe from predators.

In Europe, rabbits live underground in an intricate system of tunnels known as a warren. Since rabbits are social creatures and have very large families, many individuals will live in the same warren. The warrens have escape tunnels to ensure the rabbits can run away from any predators that find their way into the rabbit's home. Rabbits in North America do not dig deep underground tunnels. Our rabbits prefer to have shallow, sheltered burrows, which may be found within the root system of a tree, large shrub or in an area of brush. They may build a shallow burrow or nest in which they raise their young. They may abandon the burrow once the young are able to fend for themselves and then build another in a different location. You may be able to spot rabbits along densely vegetated fence rows, in piles of brush or in a sheltered garden.

Did you know?

- The star-nosed mole has 22 tentacles on the end of its nose. These strange-looking appendages are used to help the mole find its way through the darkness underground, and to locate food.
- The smallest non-flying mammal is the white-toothed pygmy shrew, with a weight of less than 2.5 grams (1/100 ounce) and a length of less than 48 mm (2 inches). This tiny creature spends much of its life underground, so it's very difficult to find.

Prairie dogs live in underground burrows

Activity

1. Distribute Student Activity Sheet #11 and help the mole find his food.

On the Ground

Mammals that walk on their toes are digitigrades (from *digit* = fingers or toes). While walking, these mammals place the majority of their weight on their toes. Hoofed animals, such as deer, moose, elk, bison, horses, cows and pigs, walk on the tips of just some of their toes.

Carnivores such as cats and small herbivores including mice and squirrels walk using a digitigrade position. These mammals use all of their digits while moving.

Omnivores such as bears, raccoons and humans place their weight on the entire foot while walking. This type of movement is called plantigrade. The tracks of these mammals are often closer together than those of the digitigrades. Although humans are primarily plantigrade, while running quickly, we switch to digitigrade.

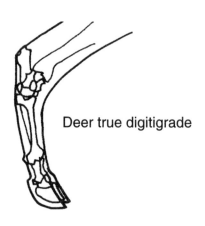

Deer true digitigrade

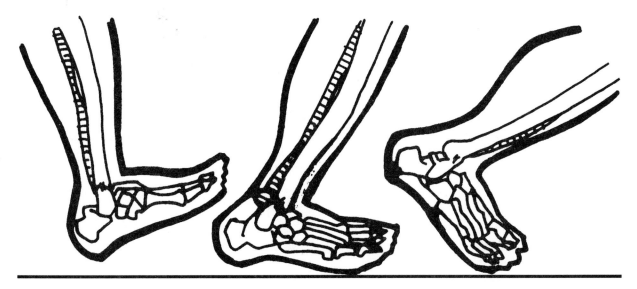

a plantigrade walking

Did you know?

- The fastest land mammal is the cheetah, which can reach speeds of almost 100 km/h (60 mph) over short distances.
- The slowest mammal is the three-toed sloth of South America, which has an average ground speed of less than 20 km/h (1/10 mph).
- The heaviest land mammal in North America is the polar bear. A male was found to weigh 900 kg (almost 1 ton) and was 3.5 m (11.5 feet) in length.

Activities

1. Find a variety of different types of music — at least 30 seconds of each type of sound. Play the music and have students move like a mammal until the music stops. The students can choose a mammal that the music reminds them of and, once the music stops, students can share the different mammals with the others in the class. Or select a mammal for each type of music and have students, as a class, try to imitate the mammal's movements.

2. *Animal Relay*

   Instruct students to line up in rows behind a designated starting line. Indicate a goal line and explain they must go to that line and back in order to touch off to the next player. The catch is that they must move to the line and back in such a way that a mammal would move. Try galloping, hopping, pouncing or "flying."

Migration

Migration is the seasonal movement of populations of animals between one region and another. This movement is usually connected with changes in season (fall to winter to spring) or the animal's breeding cycle.

Animals that migrate are called migrants. Caribou are migrants that gather in large herds to travel hundreds of kilometers across frozen landscapes to their summer grounds. Hoary bats, which travel more than 1000 km (620 miles), migrate on their own without the benefit of older members leading the way. Do people migrate? Think about migrant workers who travel to Canada for agricultural harvests, and senior citizens who travel to warmer climates to escape the cold of the Canadian winters. Can students think of any other examples?

Did you know?

• The Gray whale wins the top award for longest migration. This massive marine mammal travels from Alaska to Mexico and back, a journey that is 11,000 km (6835 miles) long and may take as long as three months.

Activity

1. Distribute Student Activity Sheet #12. Assign each student a mammal that migrates; if the class is large, put students into groups and assign one mammal to each group. Have students research the migration patterns of their species and draw a line on their map showing the mammal's migration route. Why does this mammal migrate?

    On the biggest piece of paper you can find, outline the continents of North America and South America. Cut out two silhouettes of each mammal and several copies of each mammal's tracks. Have students transfer the migration routes from their maps to the large class map by placing one mammal cut-out at the beginning of the migration route and the other at the end. Use the tracks to show the route the mammals take to migrate between these two places. Are there any similarities and differences between the migration routes? What mammals migrate the farthest? What mammal has the shortest migration route? Don't forget to also include the marine mammals that migrate. Use a dotted line instead of tracks to show their migration routes. Do all of the different mammals migrate for the same reason?

CHAPTER SIX

The Five Senses

As mammals, we depend on our five senses to function in our environment. We see with our eyes, hear with our ears, smell with our noses, taste with our mouths, and feel with our skin. Some mammals use their senses in additional ways, quite different from the way we use ours.

Bats have developed a sophisticated navigation tool. Using a series of clicks and noises, the bat is able to listen for the echo of its own voice. When the clicks the bat sends hit an object, the sound waves bounce back to the bat. The bat's brain is able to interpret these sounds into a picture of the object. This method of "seeing" in the dark is called echolocation. Bats use echolocation to not only prevent flying into things, but also to hunt for moths and other insects.

LESSON 18

Sight

You can tell a lot about a mammal by looking at its eyes. Species that are typically prey, such as rabbits and deer, must be able to see as much as possible at one time to escape danger. These mammals have their eyes located at the sides of their heads.

Predators, such as wolves and bears, have their eyes located in the front of their heads. Because of the space between their eyes, these animals see two slightly different images of the same thing. It is up to the animal's brain to put these two images together to form one picture.

The slight difference between the two images allows the animal to judge distance. This type of vision is called binocular vision.

Nocturnal animals sleep during the day and are active mainly at night. There are many reasons why an animal may be nocturnal. The bat for example, eats moths and other flying insects that mainly come out at night. Bats may also have become nocturnal in order to reduce the competition they have with birds for a similar food supply. For other mammals, being nocturnal is a means of survival. The

Prey – rabbit   Predator – wolf

darkness of night helps to hide small mammals from their predators.

Most mammals have two types of cells in their eyes, rods and cones. Cone cells help mammals see detail and colors. Rods are used for seeing under low light conditions. Nocturnal animals have mostly rod cells in their eyes. Some bats have only rods. They aren't blind, they simply cannot see color, so everything appears grey to them. People do not have as many rods in their eyes as other mammals and therefore can not see well in the dark.

When you are driving down the road at night you may see two yellow-green eyes shining back at you. Some animals have eyes that appear to glow at night. This glow comes from a mirror-like layer at the back of the animal's eye. This layer, called the tapetum, reflects light so that the total amount of light that the animal sees increases. In essence, the animal sees the light pass through its eye twice.

Did you know?

- All known species of bat are able to see perfectly well with their eyes. Bats lack a tapetum, which other nocturnal animals have, and therefore bats are not able to see well at night.

Activities

1. Illustrate the concept of binocular vision using this simple demonstration. Give each student two toilet paper rolls to look through the way they would look through binoculars. While they keep both eyes open, slowly move the ends of the rolls together until they can see only one image. Have them close the right eye. What does the left eye see? With the left eye closed note the image seen by the right eye. Students should see two slightly different images. Ask students why it might be important for a predator to see two slightly different angles of the same image. (*Answer*: the overlapping images are interpreted by the brain as a single three-dimensional picture that is important in judging distance.) Alternately, have students move the ends of the rolls outwards, towards the sides of their heads. Have them describe what they see. The two very different images are representative of what a prey species would see.

2. Try this simple experiment to help illustrate the reduced night vision of humans. Take several different colors of paper and hold them up in front of the class. Now turn off some of the lights. Randomly hold up the colors again. This time, it may be more difficult for the students to differentiate between the colors. Now try to block off as much light as possible by covering windows or closing curtains. Students will probably be unable to differentiate between the colors of paper — all will appear to be a different shade of gray. Note: if it is not possible to block light from the classroom, try this activitiy using a box that has a lid or flaps that can be opened to varying degrees and place the pieces of paper inside.

3. The glow that you sometimes see from a mammal's eyes at night comes from the mirror-like layer at the back of the eye called the tapetum. You can demonstrate the usefulness of the tapetum to mammals by conducting this simple experiment.

What you'll need:

black construction paper
aluminum foil
glue or tape
a flashlight

— Cut two small circles out of the aluminum foil to represent a mammal's eyes.
— Glue or tape the aluminum foil to a piece of construction paper. Tape the black

piece of paper and the paper with the foil "eyes" on it to a wall.

— Turn out the lights. Shine the flashlight onto the black paper.

— Now shine it on the piece of paper with the foil eyes. What differences do you see?

You were able to see the light reflected off of the aluminum foil. This reflection also happens in a mammal's eyes and increases the amount of light the mammal sees since the light is actually passing through the eye twice.

Smell and Taste

The sense of smell helps mammals detect chemicals in their environment, find food, communicate, and avoid danger. Most mammals detect smells using cavities in their heads. Generally scent detectors are located in the upper part of the nasal cavity. Mammals such as cats and mice have extra scent detectors in the roof of their mouths. These detectors are used to find a mate since they are sensitive to the scent of the opposite sex.

Scent and smelling are very important in helping mothers locate their young. In the case of mammals that congregate in large groups to give birth and raise young — such as bats, seals and sea lions — sight and sound would not be effective in distinguishing individual babies. Each young, however, has a unique scent that the mother is able to recognize and follow.

Many mammals use chemicals in order to communicate with each other. Mammals often recognize other animals by their scent. Mammals also use chemical markers to distinguish territorial boundaries and to attract a mate. Chemical messages may be more effective than vocal communication since chemicals tend to persist a long time.

The senses of smell and taste are closely linked in mammals. Have you ever had a cold and a stuffy nose? Foods taste bland when you aren't able to smell them. Mammals have taste buds on their tongues for tasting. Mammals, compared to other creatures, have an acute sense of taste and are able to use this sense to communicate, detect danger and find food. Rats have such an acute sense of taste that they are able to avoid foods that have made them ill in the past from only taking a small nibble.

Did you know?

• A wolf can smell a moose up to 2 km (1.2 miles) away.

• A deer defecates thirteen times on a winter day.

Touch

All mammals, including humans, react to different stimuli, such as heat, cold and touch. Pain is a form of touch. There has been some debate as to whether or not animals can feel pain. Since people are mammals and we feel pain, there is little reason to believe that mammals do not feel pain, at least to some extent. Studies conducted have proven that animals feel pain from being injured, as well as the type of pain associated with hunger. It is commonly believed that all mammals react to stimuli such as touch, pain, heat and cold.

If you live in an area where there are raccoons, you have probably already encountered these clever creatures trying to open the lids of garbage cans. One of the reasons why raccoons are able to open doors and latches is their incredible dexterity and their acute sense of touch. Raccoons in captivity can be seen "washing" their food. Wild raccoons do not have this behavior and have not been observed doing this. For this reason, scientists believe that the raccoon is not washing its food, but is imitating the action of catching food in the water and, possibly more importantly, using touch to recognize the object between its paws.

Did you know?

- The front paws of a raccoon have thousands more nerve endings than your hands.
- It is believed that a raccoon's sense of touch is enhanced by water.

Hearing

Most mammals have ears that protrude from their heads. The outer ear, or pinna, helps to collect and determine from which direction sound is coming. Watch your family dog, cat or rabbit — they are all able to move their pinnae. Try to move your ears. Although some people are able to wiggle their ears up and down, we cannot move our ears to help amplify sound.

All mammals have three tiny bones in their ears to help them detect sound vibrations; amphibians and reptiles have only one bone in their ears.

The sense of hearing allows animals to detect danger, communicate and hunt for food. Foxes may sit very still with their head cocked to one side staring at the ground. The fox is not resting; it is listening to mice and moles scampering through their underground tunnels. Fox use their acute hearing to hunt these tiny creatures.

Bats, whales and dolphins rely mainly on their sense of hearing to find their way around their environment. Since these mammals are often in darkness — bats are nocturnal and the ocean has low light conditions — their hearing is very acute. These mammals use a series of clicks and echoes to find their way through their environment, communicate with others and hunt (see page 35, echolocation).

Did you know?

- The loudest animal sounds are produced by Blue and Fin whales. These low-frequency pulses are 188 decibels and can be detected 850 km (528 miles) away. Compare this to a television, with a decibel range of about 70-73, and a large lawn mower of about 88 decibels.

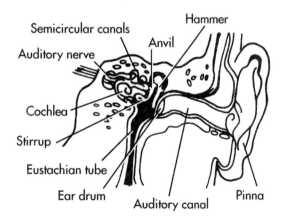

Activities

1. Go outside where there may be more noise than inside the classroom. Have students close their eyes and listen carefully. What can they hear? Now have students cup their hands behind their ears and listen carefully to the sounds around them. Can they hear more or different sounds? Are the sounds louder or softer?

   By cupping their hands behind their ears, the students are mirroring the ears that many mammals have. Mammals such as foxes and wolves have large, perky ears that they are able to move freely in order to detect very faint or quiet sounds. Why might this adaptation be advantageous to a mammal?

2. Distribute Student Activity Sheet #13. Have students match the ears to the mammals with which they belong.

3. Have students close their eyes while you play a tape of nature sounds. Include the howling of wolves, chattering of squirrels and other noises made by mammals. Ask the students how these sounds made them feel. Have them write about a time when they have heard these sound before or imagine a time and place where they may be able to hear these noises.

4. You can demonstrate the concept of echolocation using a tuning fork. As the fork is moved closer to the student's ear, the sound intensifies. The concept of sound waves can be shown using a pail of water and making ripples in it. The ripples of water move from the source outwards, just like sound waves. When the waves reach the edge of the pail, they are reflected back. Sound waves are reflected back to the bat's ears once they come in contact with an object.

CHAPTER SEVEN

Family Life

Mammals reproduce by means of sexual reproduction. A male and a female of the same species are needed for mammals to reproduce.

LESSON 22

Finding a Mate

Before mammals mate, the males and females must find each other. It is also important that the strongest and healthiest males mate with the females in order to ensure healthy offspring. For this reason, many male mammals compete with one another for the right to mate with the females. For deer and moose, this means growing antlers to fight off other, weaker males — this fighting is often called rutting. Mammals that do not have antlers fight using their claws and teeth. The winner mates with the female of his choice, the loser leaves to find another female or, in some cases, may even be killed.

Very few mammals remain together after they mate. Mammal babies are usually raised by single mothers. The male of some mammals, such as rabbits and mice, will leave of his own will, or the female will chase him away, fearing he will injure the young. Although a beaver is a rodent, like the mouse, it mates for life. In its matriarchal society, a female whose mate has died will find another mate, but the whole colony will collapse if the female dies.

In order to ensure continued success of the population, a male who is strong and healthy may mate with several different females in the same season. This practice is most noted in white-tailed deer and moose, where males must take part in the rut (competition) to win the right to mate with females.

The male fox stays with the female, helping to raise the young until the kits are grown. Wolves, although closely related to the fox, do not show the same behaviour. In wolf society, the pack is led by a dominant male and female, and it is only these two which can mate and raise young. This allows the wolves to regulate their population, based on availability of food.

Did you know?

- A female moose moans for up to 40 hours to attract a mate.
- In the spring, a groundhog does not leave his hole to predict when winter will end — he is looking for a mate!

LESSON 23

Reproduction in Mammals

Reproduction is the creation of new individuals from existing members of the population. All mammals reproduce by means of sexual reproduction. Sperm from the male must join with an egg, or eggs, from the female — this is called fertilization. After fertilization has occurred, the egg attaches itself to the uterine wall and develops into an embryo. The embryo develops in its mother's womb until it is fully developed into a tiny replica of its parents. The time it takes for the new mammal to develop in the womb of the mother is known as the gestation period. This time varies greatly between mammals and can be anywhere from 12 days (for a hamster) to 760 days (for the Asiatic elephant), depending on the species. The usual gestation period for humans is nine months.

Most mammals are placental, meaning the young grow and develop a lot while in the mother's body. Placental mammals give birth to young that are well developed and look like miniature adults. Some species, such as deer and moose, have young so well developed they are able to walk within a few minutes of birth. Why might this be important for certain types of mammals?

Other mammals, known as marsupials, carry their young in a pouch. When you think of a pouched animal, you likely think of a kangaroo or wallaby, but there is a marsupial living in Ontario. The Virginia opossum is a marsupial, the only one living in North America. Marsupials give birth to tiny, undeveloped young that must make their way to the mother's pouch. Once reaching its destination, the baby will feed on milk and continue to grow and develop until the time it is able to live on its own.

Compared to other species, mammals invest a lot of time and energy into raising their young. There are trade-offs in both having a large number of young and in looking after them for a long period of time. For example, insects and fish have a very large number of offspring but they generally do not look after them once the eggs are laid. There is usually a very high mortality rate for these species. Mammals, however, do not have as many offspring but spend a great deal of time looking after them, teaching them how to survive and defending them from predators. Mammalian young therefore have a lower mortality rate than other animals.

The number of mammary glands the female has is primarily determined by the number of young usually born. The milk that is produced by the female is rich in fat, protein, sugar, vitamins, salts and water. It is the only food her baby needs at the beginning of its life. Most babies suck milk from a nipple or teat. There are three exceptions to this, the platypus and two types of echidnas. The monotremes secrete milk from their skin and the young lick it.

Did you know?

- A baby woodchuck (also called a groundhog) is called a chuckling!
- The shortest gestation period, 12-13 days, is shared by the Virginia opossum and its rare cousin, the water opossum. The longest gestation period, of 607-760 days, is the Asiatic elephant.
- There are three mammals that lay eggs! The duck-billed platypus and two species of spiny anteaters, or echidna, are known as monotremes — mammals that do not give birth to live young but who lay eggs instead.
- Bear cubs stay with their mothers for about 17 months, learning to find food and hunt.
- A baby beaver stays with its mother for 1 to 2 years.
- A meadow vole in captivity can produce up to 17 litters of young per year.

Activities

1. Have students think about possible advantages and disadvantages to giving birth to live young compared to laying eggs.

2. Distribute Student Activity Sheet #14. Have students match babies with their species.

3. Most people have probably heard of a gaggle of geese but do you know what a group of bears or moles are called? Read on to learn the names of other groups of mammals.

Leap of leopards
Sloth of bears
Drift of hogs
Labor of moles
Shrewdness of apes
Troop of kangaroos

Pride of lions
Pod of whales
Crash of rhinoceroi
Mission of monkeys
Rag of colts

Research other creatures and find out their group names. Have students write songs, poems or stories about these mammals and their group names.

CHAPTER EIGHT

Looking for Mammals

Encountering wildlife can be the most memorable part of a walk or outing. Seeing a mammal in its natural setting can cause fear (real or imagined) or great excitement. Most of the time wild mammals are not seen; what is seen are tracks or signs left by the mammal.

Be a mammal detective. How many signs of mammals can you find in your backyard,

school ground, local park, forested woodlot, meadow, stream or pond? Take your class to some of these natural areas and have students locate as many signs of mammals as possible — make it into a scavenger hunt.

LESSON 24

Mammal Tracks

One of the most commonly seen signs that an animal is around is a track, or footprint. After closely examining the track, look for additional tracks in the same area. If there is more than one type of track, you know different animals have been here. Look at one specific set of tracks. Examining the tracks of one mammal can also help you piece together the clues to the mammal's identity.

Mammals have different types of feet and claws that would make tracks look different. They also walk differently, depending on the species. Mammals such as foxes move the front and back foot on the same side at the same time. The hind foot falls directly within the track made by the front foot, so it appears as if there is only one track. This type of walking is called direct registration and helps predators like the fox quietly stalk its prey. Although closely related to foxes, dogs walk in an indirect registration pattern. Their hind foot does not register inside the track made by the front foot.

Raccoons walk in an alternating pattern, where the front tracks and the back tracks alternate along the same line. Squirrels, rabbits and hares are known to bound or hop. Their

tracks usually consist of a complete set of four footprints in one area and another set some short distance away.

If you are near a river or stream, you may find an otter slide in the bank. Look for fresh tracks. If you find some, sit quietly and you may be lucky enough to see one of these aquatic acrobats in action.

Take a closer look at the water's edge and you may be able to see other signs of mammals. Footprints or tracks can often be seen in fresh mud. You may be able to find numerous tracks since many animals go to the water to drink and hunt. Some of the common tracks you may see around the water are made from beavers, muskrats, otters or raccoons.

Did you know?

- Aboriginal North Americans sometimes made a moccasin (boot) out of a bear's leg. They wore these shoes to disguise their footprints when neighbouring tribes were enemies.
- Mammals such as deer, mice and rabbits have regular runways they use to travel between their homes and a food supply.

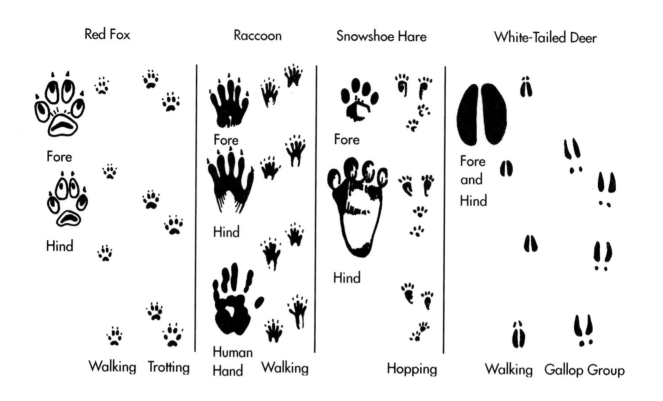

Red Fox — Fore, Hind — Walking, Trotting

Raccoon — Fore, Hind, Human Hand — Walking

Snowshoe Hare — Fore, Hind — Hopping

White-Tailed Deer — Fore and Hind — Walking, Gallop Group

Activities

1. On a warm but somewhat overcast day go outside in bare feet — make sure the area is free of debris. Fill a small pail with water. Have students dip their feet into the water and move across the sidewalk. They can walk, run, hop or leap. Look at the differences between the footprints or tracks that are made. Have other students turn their backs and let the student with wet feet choose how to move. Let the other students try to guess the movement that was done. Students can use this knowledge and apply it to a situation where mammal tracks are found. You can also make tracks using large sheets of bailing paper and by adding a little powdered tempera paint to the water to create lasting tracks when the water dries.

2. Divide students into groups of three or four. Give each team the name of a mammal (raccoon, deer, squirrel, bear, etc.). If outside, have students go to a sandy area where they can imitate the tracks of the animal. If indoors have students draw, cut out shapes of the tracks or make stamp prints by cutting a potato into the shape of the track. Once all teams have finished creating the tracks of their mammal, have the groups reassemble. One by one, the groups are to go to each set of tracks and try to identify the mammal that made them. The team with the most correct answers is the best animal detective.

3. *Plaster Casts*

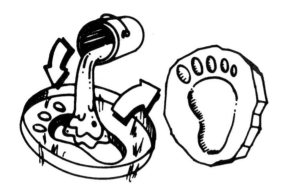

What you'll need:

cardboard approximately 30 cm x 8 cm
 (12" x 3")
stapler, tape or paperclips
plaster of Paris
a container of water
a spoon or a stick

— Look for animal tracks in the mud. See if you can find a track that is fairly deep and clear. Carefully remove any sticks or leaves that may be around the track.
— Make a ring out of the cardboard using the stapler, tape or paperclips.
— Push the ring gently into the ground so the track is in the middle of the ring. Only push the cardboard about half way into the ground.
— Mix the plaster with the water until it reaches the consistency of soft ice cream.
— Pour the mixture into the cardboard ring to a depth of about 4 cm (2 inches).
— It will take about 30 minutes for the plaster to harden, so in the meantime, see if you can find other signs of mammals.
— When you return to the track, carefully pick up the cast, cardboard and all, and gently wipe off any large pieces of dirt.
— Take the track home and use a soft cloth or old paint brush to remove the rest of the soil. Try to identify the track using the mammal tracks illustrated throughout this book or a field guide.

LESSON 25

Mammal Signs

Scat, or fecal waste, is one sure sign that an animal has been around. By examining scat, biologists are able to determine the health of the animal and the local population dynamics. Remains such as bones and hair can be found in scat and identified to learn about the diet of the animal. The size, shape and color of the scat help in identification and show important signs for skilled observers. If you find scat, be sure not to touch it unless you are with a skilled and experienced adult. Always wash your hands after handling anything that may have come in contact with an animal.

Another sign that animals have been around is the evidence left on the plant life. The presence of beavers in an area can easily be seen if you look for lodges and chewed tree stumps.

Beavers cut down trees for food and shelter using their strong teeth, leaving behind a stump with a pointed top. Although both rabbits and deer eat the tender ends of branches, you can tell which of these mammals has been eating by examining the chew marks. A rabbit's strong chiseled teeth allow it to make a clean cut through the branch. A branch that has been eaten by a deer will have a more ragged and torn end, since deer lack sharp front teeth.

Red squirrels love to feast on pine cones but they are rather messy eaters. In order to get the tender seeds inside the cone, the squirrel strips off the bracts — the coverings or casing of the seeds — leaving them and the naked cone behind. If you find little bits and pieces of

Signs that an animal has been around

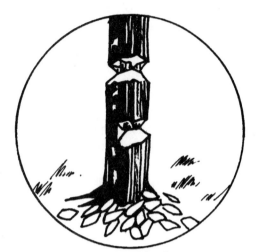

Tree gnawed by beaver

Wings of moths eaten by bat

Rabbit droppings

Squirrels strip pine cones

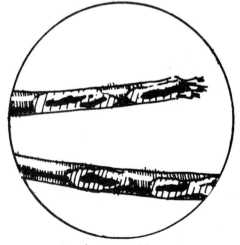

Cord gnawed by mice

Nuts split by squirrels

pine cones in a heap, you can be pretty sure that a squirrel has had dinner. Beechnuts, acorns and hickory nuts are also favorites of squirrels, and remnants of a meal in the form of the shells will often be left on rocks or fallen logs.

If you see an opening in the ground, you may have come across a burrow of a chipmunk, woodchuck, mouse, mole, weasel or skunk. You may be able to find a few holes in the same area, as many animals dig a few tunnels so that they will have an escape route. If you closely examine the hole, you may be able to determine the type of mammal that lives there. Chipmunks are neat housekeepers and have very little disruption of soil around the opening of their holes. Gophers tend to be messier builders, with mounds of soil found littering the area around the entrance. In general, the smaller the hole, the smaller the creature inhabiting the area. Mammals like to have their entrances as small as possible to help keep predators out.

Did you know?

- A pile of moth or insect wings is a sign of bats. Some bats eat hundreds of moths each night. Preferring only the tender body, the bat strips off the wings, allowing them to fall to the ground.
- The new antlers of deer family are covered with "velvet." Once the antlers have fully developed, this covering begins to itch. The deer rubs its antlers against bushes and tree trunks. Strips of antler velvet can often be seen hanging from low branches. One side has a fuzzy feel while the other is smooth.

Activity

1. Collect natural materials — sticks, leaves, stones, soil, feathers, seeds and cones. After conducting some research, have students build a habitat for a mammal. This activity may be easier to do in small groups. Some habitats to consider: squirrel nest (also called a drey), beaver dam or a mole burrow.

CHAPTER NINE

Mammals and People

Everyone knows about dinosaurs — creatures that once lived on Earth but are no longer found here. Extinction is a natural process whereby some species die and others take their places in the food chain. But the way we live our lives may affect animals in many ways, and sometimes puts them in danger.

LESSON 26

Mammals in the City

Cities are prime habitat for mammals because there tends to be fewer predators there than in the country. Hunting is not allowed in cities, thus reducing the chance of being caught. The city tends to be warmer than the country with milder winter weather. There is an almost endless supply of food and water from stores and people's garbage and litter. All of these human-made conditions make for a very hospitable environment for many different types of mammals, including rabbits, mice, rats, squirrels, and skunks. At the same time, automobile traffic presents a hazard to mammals living in these constructed environments.

Toronto has been called the raccoon capital of North America. These dynamic animals have found an abundant food supply and shelter in this city and their population has flourished as a result. It is not uncommon to see raccoons in parks, cemeteries and treed neighborhoods. They take advantage of drainage pipes and culverts and use these "highways" to travel between corridors of trees and bushes. The only predator of raccoons that is also found in the city are coyotes.

Coyotes are able to live in cemeteries, golf courses and ravines within city limits. These mammals are relatives of dogs. Coyotes find more than enough prey in cities, where they can catch raccoons, squirrels, mice, rats, small dogs and cats, rabbits, skunks, birds, reptiles, amphibians, fish and insects. The coyote's ability to feed on a large number of creatures and its nocturnal behavior allows its population to thrive within cities. Despite their relatively large numbers in some Canadian cities, coyotes are seldom seen by people.

There are different ways in which people can interact with mammals. Consumptive uses include those that kill mammals or remove them from their natural habitat in the wild. Hunting reduces the population of a species, but this consumptive use is not considered the primary loss of wildlife species. In Canada, loss or degradation of natural habitat — due to air and water pollution, draining wetland ecosystems, soil erosion, deforestation and urban expansion — is reducing numbers of wildlife species.

A variety of non-consumptive uses of mammals are becoming more popular with people. You can enjoy mammals by watching them, photographing them and drawing them in their natural settings.

Did you know?

- Although most rats are no longer than 30 cm (12 inches) from nose to tail, they are able to leap the length of a small sofa and the height of a table. They can also squeeze through openings no larger than a quarter.

Activities

1. How many "famous" mammals can students think of and what kind of image does each represent (e.g., Smokey the Bear, Ranger Rick the raccoon)? Think of sports teams that are named for mammals — the Florida Panthers, BC Lions, Hamilton Tiger Cats, Phoenix Coyotes, Vancouver Grizzlies. Which part of the food chain did these teams choose? Why? What companies use mammals as their logo? List as many as possible (e.g., World Wildlife Fund, Wildlands League, ESSO, MGM Studios, etc). Why have these companies chosen mammals to represent them?

2. Have students collect advertisements from magazines or articles from the newspaper that show or talk about mammals. How are mammals portrayed in each? What type of message are the advertisers trying to show in each of the ads? Do you think mammals help to sell products? Create an advertisement for a product and use a mammal as a spokesperson. Explain the reason why you chose that mammal for the product (e.g., a skunk for a room freshener).

3. Have students write a story or draw a picture explaining what wilderness means to them. Similarly, do the same lesson but have them explain what they think wilderness means to mammals. Talk about how their stories or pictures are the same and how they are different.

LESSON 27

Mammals in Jeopardy

Each year the Committee on the Status of Endangered Wildlife in Canada (COSEWIC) reviews the populations of wildlife in Canada and, if needed, assigns an official designation. There are five possible designations for a species in decline: extinct, extirpated, endangered, threatened or vulnerable.

Extinct: the species no longer exists anywhere in the world.

Extirpated: the species is no longer found in its original habitat but may be found in other places.

Endangered: the species faces extirpation or extinction.

Threatened: the species is likely to become endangered if something is not done to help the species recover.

Vulnerable: the species is sensitive to the impact of human activities (for example, the species requires very specialized habitat in which to live and that habitat is being destroyed) but is not yet considered threatened or endangered.

There are some species of mammal that were once found in Canada but which no longer live here. Some of these species have been forced to move elsewhere in search for food or shelter. But others cease to exist at all. The Sea Mink, the Queen Charlotte population of the Woodland Caribou, and the Eastern Elk are extinct. These species were once indigenous to Canada but are no longer found anywhere in the world. There are many more mammals that face this fate. It's not too late to help save these species. People who care about mammals can make a difference.

In the past there were many reasons for animal extinction. Today, loss or degradation of habitat is the main reason creatures in North America are at risk. Across the continent, forests have been clear cut, prairie ecosystems have been destroyed and wetlands have been drained and filled. Acid precipitation, pesticide use, improper handling and disposal of hazardous and municipal wastes, poor soil management practices, urban expansion and industrial development also lead to the destruction of wildlife habitat. This alteration of habitats leaves thousands of animals without suitable habitat in which they can live, find food and reproduce. If animals are unable to adapt to these new surroundings, they must either move to new areas or they will die.

Activities

1. Have students find out what is being done in Canada and their province to preserve wildlife and their habitats. Consider looking at the Provincial Parks Act, Public Lands Act, Wilderness Areas Act, Ontario Endangered Species Act, Fish and Wildlife Conservation Act, National Parks Act and the Canadian Wildlife Act, to name just a few.

2. Create a wildlife refuge in your schoolyard and watch to see what creatures come to take a look. To attract small animals such as mice and chipmunks, create a brush pile of leaves, branches and twigs. This habitat will provide the animals with safety from predators. Build bat houses to attract these winged mammals!

3. A game of musical chairs can be used to illustrate the effects of shrinking habitats on wildlife, a process potentially leading to extinction. Set up chairs in a circle, one per student. Tape a picture of a dinosaur to one of the chairs to represent extinction. Have students walk around the chairs to music or environmental sound effects. When the music stops, each student must find a chair to sit in. The child in the dinosaur chair leaves the game and a chair is removed (note: the chair of extinction is never removed from the game). For a twist, each student can be given the name of an animal or species that they are to represent. As the students are forced to sit in the dinosaur chair, they become extinct and so too does that species of animal. A discussion can occur each time a species becomes extinct.

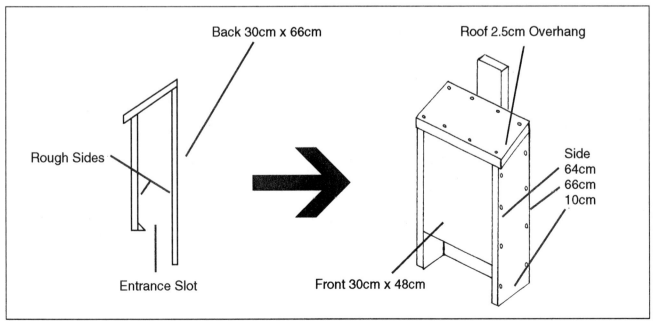

Bat box

Focus on the other creatures that would be affected in the food chain.

4. Give students a voice! After talking about why some mammals are at risk, have students discuss reasons why these mammals and their habitats should be saved. As a class, brainstorm ways of getting the message of species and habitat preservation out to others. Have students create a display for the school, local community centre or library. Draw pictures or write letters to the Minister of the Environment, Minister of Natural Resources and their local Member of Provincial Parliament telling them why species should be preserved. Try to get as many people in the school involved as possible!

Student Activity Sheets

Spot the Mammal

dinosaur

fox

penguin

seal

cicada

shark

bat

Rhyming Animals

mole	—	vole
hare	—	_____
bat	—	_____
dog	—	_____
mink	—	_____
	—	_____
_____	—	_____
_____	—	_____
	—	_____

What Makes a Mammal a Mammal?

tail

hibernate

scales

eat meat

hair

feathers

migration

backbone

fur

Pin the Tail on the Mammal

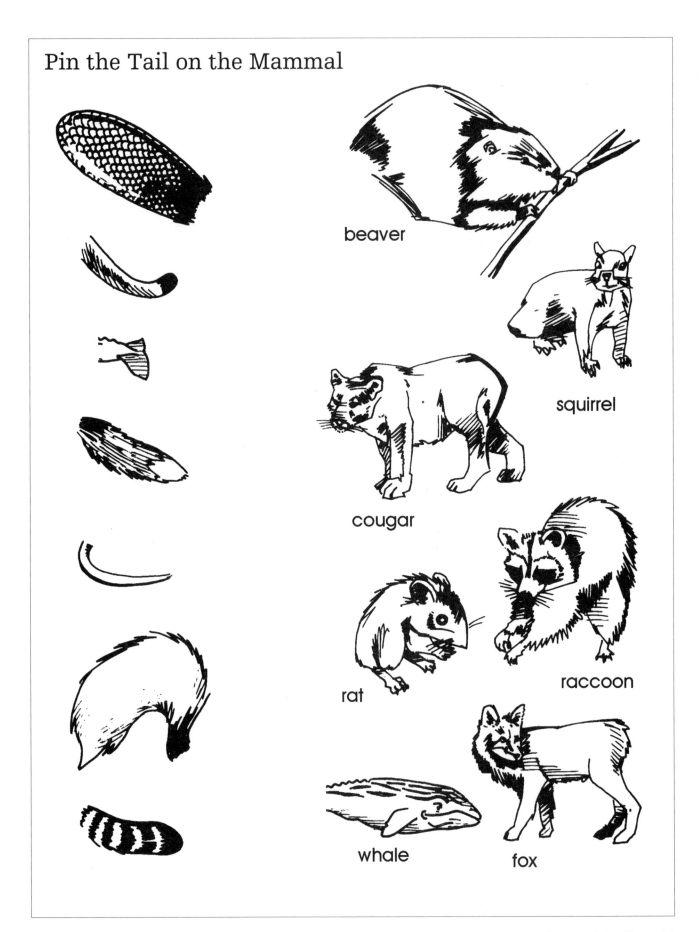

beaver

squirrel

cougar

rat

raccoon

whale

fox

Build a Skeleton

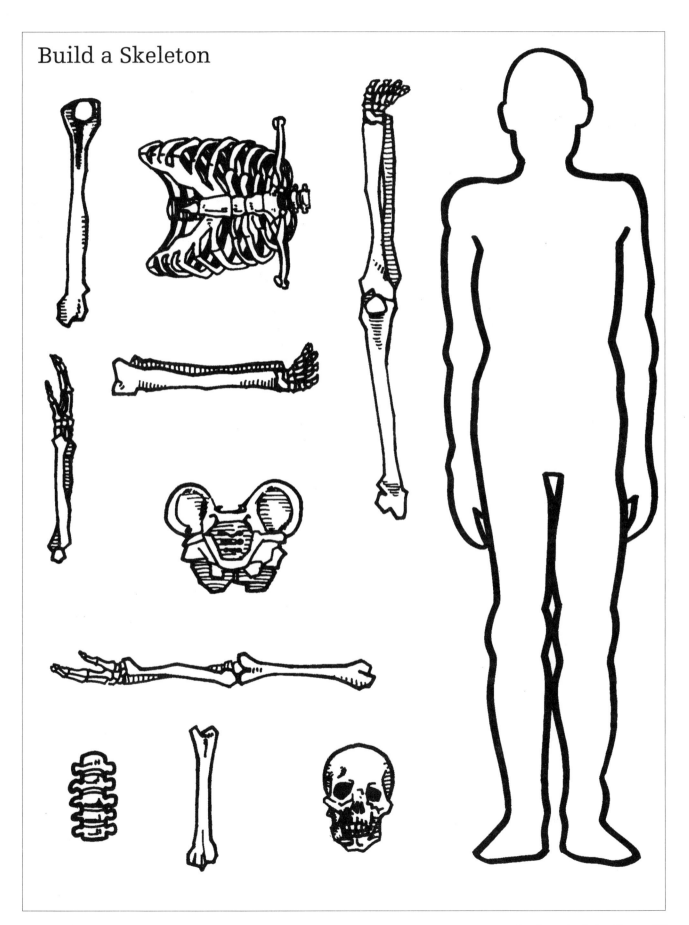

Making a Food Chain

Food Chains:

plant → plant-eater → insectivore (or carnivore) → carnivore (or top carnivore) → top carnivore → decomposer

Prairie/Grassland

small white lady's slipper → grasshopper → Eastern kingbird → blue racer snake → badger → worm

Wetland

plankton → aquatic insect (water boatman, mosquito larvae) → small fish → water shrew → otter → snail

Urban

dandelion → earwig → American toad → garter snake → red fox → carrion beetle

Woodland

small woodland plant (trout lily, trillium, young aster leaves) → caterpillar → young bird (on nest) → red squirrel → Canada lynx → slug

Aquatic–marine

plankton → small fish → cod → harp seal → polar bear → marine bacteria

Hide and Seek

Teeth Match

Food

deer

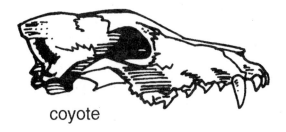

coyote

mole

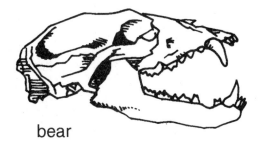

bear

wolf

Count the Quills

Whale Bones

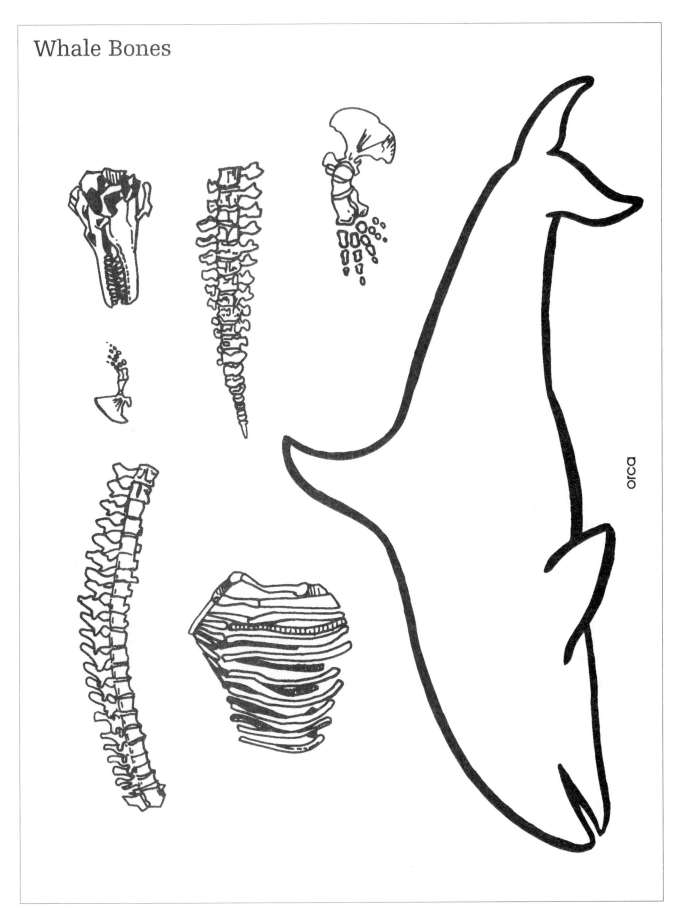

orca

Mole Maze

Migration Map

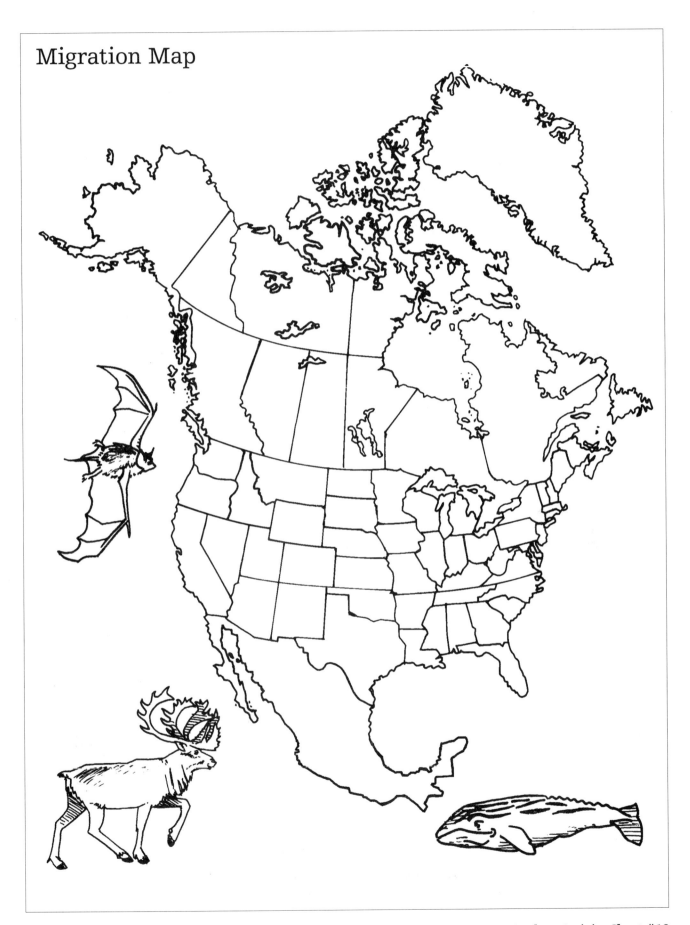

Ear Match

raccoon

lynx

deer

bat

Baby Animal Match

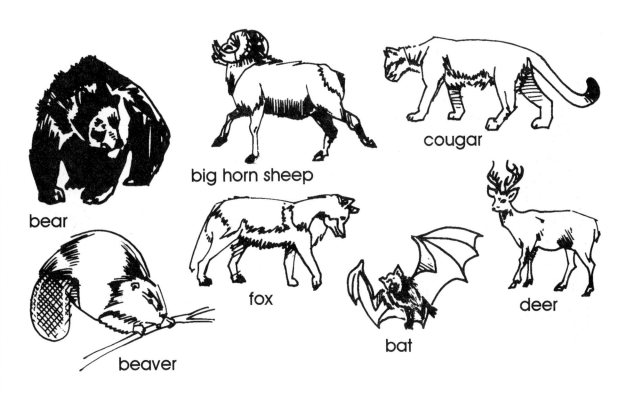

bear

big horn sheep

cougar

fox

bat

deer

beaver

Fact Sheets

Common Shrew

Scientific name: *Sorex cinereus*
Order: *Insectivora* — shrews and moles

Look at Me
- brownish-black with a silver-grey underside
- weight between 3.5 and 5.5 g (less than 1/5 oz)
- 7.5 to 10 cm (3-4 inches) long from nose to tail; tail between 3 and 4.5 cm (1.2-1.7 inches) long

Find Me
- the most widely distributed shrew in Canada
- lives in open meadows, marshes, bogs, and forests
- does not build underground burrows but often will use those abandoned by other animals
- active throughout the day and night

Let's Eat
- voracious hunters, can see their prey more than 25 feet away; but they usually hunt by smell
- enjoys moth and beetle larvae, caterpillars, crickets, worms, centipedes, spiders, slugs, seeds, and occasionally young mice and small amphibians such as salamanders
- plant material makes up about only 1% of total diet; may be accidentally ingested

My Enemies
- hawks, owls, shrikes, herons, mergansers, foxes, coyotes, weasels, snakes, fish, and other shrews

Family Life
- in northern range, shrews have one litter per year; in the southern areas, they may have several
- gestation period about 18 days
- between 2 and 10 babies born in each litter
- babies weigh 0.25 g (less than 1/100 oz) at birth; are blind, deaf, toothless, and furless
- young follow the mother in a single-file line
- lifespan about 18 months

I'm Special
- must eat their own body weight in food every day to sustain high metabolic rate and, if deprived of food, may die within a few hours

P.S.
- the common shrew is also known as the "masked shrew"

Words to Learn
burrow
insectivore or Insectivora
life-span

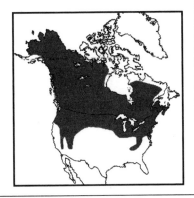

Fact Sheet

Little Brown Bat

Scientific name: *Myotis lucifugus*
Order: *Chiroptera* — bats

Look at Me
- rusty-brown with a greyish underside
- weight about 7 g (1/4 oz); approximately 10 cm (4 inches) long
- can have a wingspan of nearly 25 cm (10 inches)
- can fly at speeds between 9.5 and 16 km/h (6-10 mph)

Find Me
- bats are nocturnal: they are less active in the day and are mainly found at night
- during the day, you may find bats sleeping in caves, mines, culverts, attics, or hollow trees, behind shutters or under roof shingles
- well adapted to life in both country and city

Let's Eat
- these bats eat mosquitoes, mayflies, beetles, and other flying insects —- and are particularly fond of moths
- one bat can eat up to 600 mosquitoes in one hour
- although bats can see perfectly well during the day, their eyesight is limited at night. To compensate for this lack of vision, bats have developed a special sense called echolocation. They use this sense to find their way in the dark and to hunt for food.

My Enemies
- owls, snakes, raccoons, hawks, shrews, fish, domestic cats and dogs, and even other bats

Family Life
- gestation period between 50 and 60 days
- bats give birth to one baby at a time, usually in the late spring or early summer
- born furless, unable to fly; with eyes closed
- at birth, babies weigh about 2.5 g (less than 1/100 oz) and are about 5 cm (2 inches) in length
- each baby bat has a unique scent and call; this allows a mother to find her baby among the thousands in a colony
- little brown bats can live 25 to 30 years in the wild

I'm Special
- bats are the only flying mammal on Earth
- bats do not fly with wings; they fly with their hands. The "wing" is actually a thin layer of strong skin stretched over long fingers. "Chiroptera" means "hand-wing."
- in the late fall, dozens of bats congregate in a hibernaculum, find a place to hang, and wrap their wings around their bodies to keep warm

P.S.
- there are nearly 1000 different kinds of bats in the world. They make up over 1/4 of all mammals on Earth
- bats have lived on Earth for nearly 50 million years

Words to Learn
Chiroptera
echolocation
hibernaculum
nocturnal

Snowshoe Hare

Scientific name: *Lepus americanus*
Order: *Lagomorpha* — rabbits, hares and pikas

Look at Me
– rusty grey-brown in summer; coat turns silky white in winter, with only the tips of the ears staying black.
– adults reach an average size of 35-50 cm (14-20 inches) from nose to tail and can weigh between 1 and 2 kg (2-4.5 pounds)

Find Me
– found throughout the northern climates in Canada and Alaska
– the hare's habitat includes forested regions, swamps, thickets, and vegetated riversides

Let's Eat
– during the summer, hares eat grasses, herbs, buds, fruit, strawberries, dandelions, clover, seeds, roots, and leaves
– in the winter, when food is sparse, the snowshoe hare eats the bark of trees, pine needles, thin branches, twigs, seeds, and pine cones
– is crepuscular, which means it feeds mainly at dawn and dusk

My Enemies
– eagles, hawks, ravens, foxes, wolves, great horned owl, coyotes, mink, lynxes, and bobcats
– hares may freeze in one spot or run from predators

Family Life
– after a 36-day gestation period, babies are born in May
– females give birth to 1 to 9 young
– in their first year of life, babies are called "leverets"
– a female can have up to 4 litters each year

I'm Special
– the snowshoe hare gets its name from its wide hind feet. They help the rabbit walk around its snowy habitat
– the ability to change colour from brown in summer to white in winter gives this hare its other name, "varying hare"

P.S.
– hares communicate with each other primarily through the thumping of their hind feet against the ground
– some ways to tell rabbits from hares:

Hares	Rabbits
Ears: long with black tips	*Ears*: shorter, no black tips
Babies: furred, eyes open, ready to leave nest	*Babies*: hairless, blind, helpless
Group life: loners, loose groups	*Group life:* strictly organized groups

Words to Learn
camouflage
crepuscular
Lagomorpha
leverets

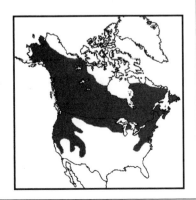

Gray Squirrel

Scientific name: *Sciurus carolinensis*
Order: *Rodentia* — rodents

Look at Me
- most have grizzled-grey fur with a yellowish-brown underpart
- some gray squirrels, especially those around the Great Lakes, can be black in colour. These are often referred to as black squirrels
- weight between 350 and 700 g (12-25 oz); length 35 to 50 cm (14-20 inches). Half of the length is tail.
- its long, bushy tail can be used as a rudder while swimming, an umbrella if it is raining, a blanket to keep warm; a squirrel fluffs it as a sign of anger, waves it during courtship, flicks it to greet others, and uses it as a signal to warn others of danger.

Find Me
- arboreal, which means that they spend most of their time in trees
- found living in eastern hardwood or mixed forests, they prefer deep forest over edge habitats
- gray squirrels are commonly found in urban parks where there are fewer predators and an abundance of food

Let's Eat
- gray squirrels are diurnal, meaning they are active during the day
- gnawing teeth called incisors continually grow in all rodents; in squirrels, these teeth are found in both the upper and lower jaws
- although acorns are their favourite, gray squirrels will also eat walnuts, beechnuts, pine seeds, hazelnuts, wild grapes, blueberries, cherries, insects, flowers, mushrooms, and even bird's eggs
- can be found nibbling on deer antlers, bones, and turtle shells, which are all sources of calcium

My Enemies
- coyotes, foxes, weasels, mink, raccoons, skunks, bobcats, hawks, owls, snakes, and domestic cats and dogs

Family Life
- after a 44-day gestation period, 1 to 6 young are born in a cozy den or nest
- most squirrels have 2 litters each year, one in March and the other in July
- at birth, babies weigh 15 -18 g (about 1/2 oz) and are less than 12 cm (5 inches) long
- most gray squirrels live less than 2 years, but they can live as long as 15 years, and even longer in captivity

I'm Special
- squirrels always come down a tree head-first and run up the tree head-first so they can look for danger
- gray squirrels grow fur on their feet during the winter to keep their feet warm and to help them climb slippery branches

P.S.
- the Latin name for the squirrel means "shade tail"

Words to Learn
arboreal
diurnal
incisors
Rodentia

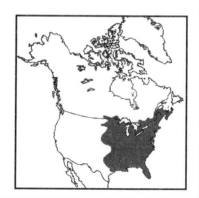

Porcupine

Scientific name: *Erethizon dorsatum*
Order: *Rodentia* — rodents

Look at Me
- weighs up to 15 kg (33 pounds), and is almost 1 m (3 feet) long
- quills found on the face just over 1 cm (less than 1/2 inch) long, while quills on the back can be up to 12.5 cm (5 inches) in length
- no quills on legs or underparts

Find Me:
- found across Canada and into a large portion of the USA, in a variety of vegetation types from semi-desert to forests to tundra
- can be found high in trees such as aspen, hemlock, and pine
- they are nocturnal, active mostly at night; they can sometimes be seen sleeping in trees during the day

Let's Eat
- this rodent loves the tender inner bark of trees and branches
- a herbivore that eats twigs, branches, buds, leaves, grass, shoots, roots, clover, and flowers such as roses, violets, and dandelion
- with poor eyesight, porcupines use their sense of smell to find food

My Enemies
- cougars, Canada lynxes, bobcats, wolverines, wolves, red foxes, bears, fishers, and great horned owls
- when a porcupine is threatened, it will shake its body to rattle its quills, growl, stamp its feet, chatter its teeth, and, if needed, swat its attacker with its tail

Family Life
- mating occurs in September and October; a single baby is born between March and May
- baby porcupines are called porcupettes
- babies are born with their eyes open; they weigh about 0.5 kg (1 pound) and are about 30 cm (12 inches) long

- the young are born with soft quills. It takes about an hour for the quills to harden and sharpen.
- they are mostly solitary animals, but might congregate in winter to share a food source

I'm Special
- unlike squirrels, porcupines come down a tree tail first
- the end of each quill has a barb, like a fishing hook, which makes removal from an enemy very difficult

P.S.
- porcupines love salt. They will eat anything salty, including wooden canoe paddles that have perspiration on them.

Words to Learn
barb
herbivore
porcupette
quill
Rodentia

Beluga

Scientific name: Delphinaptrus leucas
Order: *Cetacea* — whales, dolphins, and porpoises

Look at Me
- this small, toothed whale is completely white
- adults weigh about 1.5 tons and are nearly 5 m (16 feet) in length
- the bump on the top of a beluga's head is believed to contain a special organ that is used for echolocation
- belugas of the St. Lawrence population are generally larger than those of the Hudson Bay population

Find Me
- these whales mainly live in shallow Arctic waters in northern Canada; however, a small population lives in the waters of the St. Lawrence
- there are 3 main populations of beluga whales in and around Canada: near Baffin Island, Davis Strait, and Ungava Bay

Let's Eat
- they have 16 to 20 small, flat teeth that are good for grabbing prey but not for tearing flesh; therefore, they usually eat their food whole
- belugas enjoy a wide variety of fish, such as salmon, capelin, cisco, flounder, pike, char, and cod; they will also eat squid, octopus and shrimp
- using their flexible necks, they are able to "sweep" the bottom of the seabed for worms, mollusks, and small fish

My Enemies
- this small whale is known to be hunted by orcas and humans

Family Life
- belugas mate in May, and have a gestation period lasting 14.5 months
- populations living near Greenland have calves in March; while the Hudson Bay population gives birth in August
- one calf is born to a female every 2 years; the baby will stay with its mother, nursing off of her rich milk, for up to 2 years
- a newborn weighs almost 80 kg (175 pounds) and is 1.5 m (5 feet) long

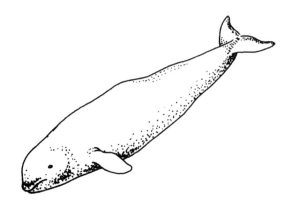

- a calf is greyish-brown when it is born and does not turn completely white until reaching the age of 6 or 7 years
- these social whales travel in pods (family groups)
- may live up to 30 years

I'm Special
- belugas have been called "canaries of the sea" because of their clicking and chirping sounds that can be mistaken for songbirds

P.S.
- "beluga" means "white one" in Russian

Words to Learn
blubber
Cetacea
echolocation

Black Bear

Scientific name: *Ursus americanus*
Order: *Carnivora* — meat eaters

Look at Me
- the black bear is the smallest bear in North America
- height at shoulder is between 90 and 110 cm (35-45 inches)
- adults can weigh between 55 and 225 kg (220-500 pounds)
- despite their name, they are not all black — they can be brown or even white

Find Me
- they prefer thickly wooded areas that are near a creek, stream, or lake
- you can find them "dog-paddling" in water

Let's Eat
- will eat just about anything: carrion, ants, grasshoppers, acorns, grasses, roots, blueberries, strawberries, apples, honey, nuts, fish, small mammals, or birds
- black bears rarely kill an animal for food unless it is easy prey

My Enemies
- black bears are able to quickly climb trees in order to escape danger or to get a better look at their surroundings
- cougars, lynxes, grizzly bears, and adult black bears are known to attack young or sick black bears

Family Life
- in late fall, females find a den in a cave, hollow log, or old stump for their winter sleep, and give birth (once every other year) in January or February.
- black bears usually have 2 cubs at a time
- cubs weigh about 230 g (8 oz) when born
- cubs leave the den for the first time in mid-April, when they have reached the size of puppies
- females are called sows; males are called boars
- a black bear can live 10 to 15 years in the wild

I'm Special
- black bears can sprint up to 55 km/h
- black bears have a good sense of smell and hearing but poor eyesight; they are also color blind and can only see in shades of grey

P.S.
- Famous black bears: Smokey the Bear, who was saved from a forest fire by a forest ranger and became a symbol in the US for forest-fire prevention; and Winnie, a female black bear from White River, Ontario, that was the inspiration for A.A. Milne's Winnie the Pooh.

Words to Learn
Carnivora

Canada Lynx

Scientific name: *Lynx canadensis*
Order: *Carnivora* — meat eaters

Look at Me
- can weigh between 8 and 11 kg (17-25 pounds)
- can easily be identified by its short, black tail and ear tufts
- in summer the fur is reddish-brown; in winter it changes to light grey
- its exceptionally long back legs help it jump large distances

Find Me
- in old-growth boreal forest where there is thick undergrowth
- it is mainly nocturnal and rarely seen during the day

Let's Eat
- main diet during the winter is snowshoe hares; it eats up to 200 in a year
- during the summer it will eat hares, as well as grouse, mice, voles, squirrels, and foxes
- like other wild cats, they hunt in 4 main steps: hiding, stalking, pouncing, killing
- a lynx might hide uneaten food for later

Family Life
- except when raising young, adult lynx are solitary
- mating occurs in February and March; 2-4 kittens are born in April or May
- sometimes use underground dens, but prefer to raise young in hollow logs or piles of brush
- the babies usually leave their mother when they are less than a year old

I'm Special
- this cat's paws act like snowshoes; by spreading its toes, the lynx can walk on top of the snow
- unlike other mammals, lynx and other members of the feline family can retract their claws

P.S.
- the lynx is sometimes called the "ghost of the forest" because it is so difficult to see
- just like your pet cat, wild kittens purr
- although lynxes are very good climbers, you will rarely see one in a tree

Words to Learn
agile
Carnivora
feline
nocturnal
retractable

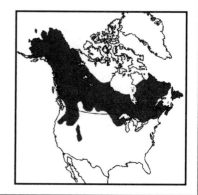

White-Tailed Deer

Scientific name: Odocoileus virginianus
Order: *Artiodactyla* — cloven-hoofed mammals

Look at Me
- white-tailed deer get their name from the 30-cm (12-inch) tail that is white on the underside
- they stand about 1 m (3 feet) high at the shoulder and can weigh over 90 kg (200 pounds); a large buck can weigh as much as 180 kg (400 pounds)
- adults are reddish-brown in summer and greyish-brown in winter

Find Me
- prefers to live in the shelter of woodlands and at the edges of forests, and along the edges of creeks
- when there is a heavy winter snow, you can find dozens or even hundreds of deer congregated in stands of evergreen trees — these areas are called deer yards

Let's Eat
- their front teeth are sharp and pointed outwards for nipping and pulling leaves
- they prefer leaves, grasses, and buds, but will also eat shoots, bark, nuts, fruit, mushrooms, lichen, twigs, clover, berries, apples, acorns, cedar, water lilies, evergreen needles, and spruce/white pine branches in the winter
- deer chew their cud like cows
- deer feed mainly at dawn and dusk when there are fewer predators

My Enemies
- foxes, dogs, wolves, grizzly bears, mountain lions, eagles, coyotes, bobcats — and humans
- when danger is near, deer will stamp their feet and raise their tails in order to warn other members of the herd

Family Life
- white-tailed deer often have twins or even triplets in the late spring, on a bed of dried leaves and grasses on the forest floor
- babies weigh 2.5 kg (5.5 pounds) each at birth, are able to stand within 30 minutes
- fawns have a light-brown spotted coat that helps them blend in with their surroundings
- fawns lose their spots after about 3-1/2 months

I'm Special
- white-tailed deer can run at speeds up to 60 km/h (37 mph) and can leap obstacles as high as 2.5 m (8 feet)
- deer are the only creatures to have true antlers. Antlers are only temporary; they are grown and shed each year by the bucks. The size and the number of points in antlers relate to the quality of the animal's diet and its age.

P.S.
- males or bucks fight, using their antlers in order to win a female; this is called a rut. Sometimes the antlers become entangled and, if the two bucks aren't able to break free, they will starve to death.

Words to Learn
antlers
Artiodactyla
camouflage
dewclaws
rumen
rut

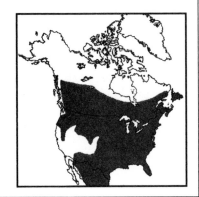

Virginia Opossum

Scientific name: *Didelphis virginiana*
Order: *Marsupialis* — marsupials

Look at Me
- greyish-white with rat-like features
- about the size of an average house cat: weighs between 2 and 7 kg (4.5-15 pounds); approximately 90 cm (36 inches) long
- has a long, pointed pink nose and a rat-like tail

Find Me
- found in southern portions of Canada; the largest Canadian population lives in southern Ontario
- prefers woodland forest or brush near streams or swamps
- can also be found around farmland or in wooded parts of cities
- nocturnal
- does not hibernate, but activity is reduced in the cold parts of winter

Let's Eat
- this omnivore will eat just about anything: fruit, vegetables, eggs, beetles, ants, worms, snakes; might also kill and eat farmer's chickens
- cannot see well, so uses sense of smell to find food
- they also eat road kill, and are often struck on the highway as a result

My Enemies
- known to "play dead" when confronted by an enemy, and can remain motionless for up to a few hours. Some scientists believe that it is not pretending to be dead, but has fainted from fear.

Family Life
- 2 litters born each year and each litter has about 20 babies
- 20 baby opossums could fit into a tablespoon
- young are born in leaf-lined dens found in holes and hollows of trees, or in rock crevices
- gestation period between 12 and 13 days
- babies must crawl up to the mother's pouch where they stay for several weeks, drinking milk

- in 10 weeks the young can venture out of the pouch

- most opossums live about 2 years

I'm Special
- opossums are the only marsupials found in Canada

P.S.
- the opossum's hairless tail is susceptible to frostbite; many opossums are missing parts of their tails
- the name comes from the Algonquian word "apasum" which means "white animal"

Words to Learn
gestation
marsupial
pouch

Red Fox

Scientific name: *Vulpes vulpes*
Order: *Carnivora* — meat eaters

Look at Me
– between 90 and 115 cm (35-45 inches) in length; a third of that is tail
– the males are slightly larger than the females
– most weigh between 3.5 and 7 kg (7-15 pounds)
– its paws, ears, and muzzle are tinged with black
– can range in color from red to brown to mostly black to pure black to pure white

Find Me
– one of Canada's most wide-spread mammals
– are often found in urban areas near streams and heavily treed areas, such as ravines, parks, and cemeteries
– they are nocturnal, hunting from dusk to dawn, under the cover of night

Let's Eat
– they mainly eat small mammals such as voles, mice, lemmings, squirrel, hares, and rabbits; have been known to eat other creatures such as beavers, reptiles, birds, fish, and insects
– foxes will also eat berries and plants when they are available; when food is scarce, they eat whatever is available, including garbage
– has a keen sense of smell, and can distinguish scents from 1.5 km (1 mile) away

My Enemies
– wolves, coyotes, domesticated dogs, bobcats, lynxes, and cougars

Family Life
– a fox's home range is about 5 km² (2 square miles)
– they breed in late December and mid-March
– after mating, the male and female look for a cozy den in an abandoned burrow, cave, hollow log, or dense bushes along fencerows
– pups are born between mid-March and May
– up to 5 pups are born naked and blind, weighing less than 115 g (4 oz) each
– when the pups are able to feed themselves, they leave the parents' den, usually when they are 3 months old
– foxes live alone unless they are raising young
– they can live up to 12 years in the wild

I'm Special
– foxes hunt much like cats, crouching low to the ground and then pouncing on prey
– in winter, stiff hairs grow between a fox's toes, helping it balance on ice and keep its feet warm

P.S.
– a fox keeps warm while sleeping in the winter by covering its nose and head with its bushy tail
– a male fox is called a dog; a female is called a vixen

Words to Learn
Carnivora
muzzle
nocturnal
predator
vixen

Harp Seal

Scientific name: *Pagophilus groenlandicus*
Order: *Pinnipedia* — seals and sea lions

Look at Me
- both the male and female adults are about the same size: 2 m (6.5 feet) long, and between 135 and 180 kg (300-400 pounds) in weight
- the male is silvery-grey with a black head and a horseshoe-shaped marking running along its back; it is this pattern that gives the harp seal it's name
- females have similar markings but are usually lighter in color than the males

Find Me
- in the North Atlantic, just below the Arctic circle, near Newfoundland and in Hudson's Bay
- in the summer these seals live in the eastern Arctic waters; in the winter they migrate to the Gulf of St. Lawrence and to the coast of Nova Scotia

Let's Eat
- seals have teeth, but they do not chew their food; they swallow it whole
- harp seals enjoy mackerel, herring, cod, and other fish, shrimp, crabs, and octopus

My Enemies
- polar bears, orcas, sharks, and humans

Family Life
- the gestation period for seals is 12 months
- newborn seals, called pups, weigh about 5.5 kg (12 pounds) and are about 80 cm (31 inches) long
- pups must be born out of the water
- each baby seal has its own unique scent that the mother follows
- babies are covered with fluffy white fur, and so are sometimes called "whitecoats"
- the young nurse for 10 to 14 days; then they can fend for themselves. The pup then lives off of its stored blubber and loses its fluffy white coat.
- seals can live for 25 years in the wild

I'm Special
- spends 4-6 weeks of the year on ice floes; the rest of the time, seals are in the water
- throughout their lives harp seals will travel more than 120,000 km (74,000 miles)
- this seal is able to stay underwater for up to 15 minutes; while underwater, its metabolism drops and its heartbeat slows from 150 times per minute to about 10 times per minute

P.S.
- its scientific name translates to "the ice lover from Greenland"
- this seal is also known as the "saddleback seal" because of its black markings
- hunters prize the fluffy, soft, white coat of newborn pups. Because of this, it is believed that the population fell from over 2,500,000 individuals to fewer than 1,000,000.

Words to Learn
blubber
metabolism
Pinnipedia

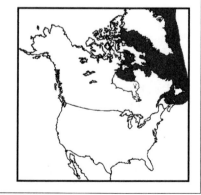